THE WORLD NATURALIST / Editor: Richard Carrington

The Natural History of Viruses

C. H. Andrewes

W · W · NORTON & COMPANY · INC ·

New York

PRINTED IN THE UNITED STATES OF AMERICA

3 4 5 6 7 8 9 0

Contents

Part 1 Viruses in General

1 INTRODUCTORY   3

2 THE NATURE AND VARIETY OF VIRUSES   11

3 THE VIRUS AND ITS HOST   24

4 DEFENCE AGAINST VIRUSES   34

Part 2 Transmission of Virus Infections

5 CATCHING COLDS   41

6 INFLUENZA, RINGER OF CHANGES   54

7 POLIOVIRUS AND ITS RELATIONS   68

8 MOSQUITOES, HORSES, BIRDS AND SNAKES   77

9 MOSQUITOES, HERONS AND PIGS   87

10 DENGUE: A CHANGING TROPICAL DISEASE   93

11 YELLOW JACK   98

12 TICKS   107

13 SAND-FLIES AND MIDGES   113

14 ALTERNATIVE ROUTES   119

15 APHIDS, HOPPERS AND BEETLES   125

16 BOOTS, WORMS AND FUNGI   133

Part 3 Other Aspects of Ecology

17 RABIES AND MORE ABOUT RESERVOIRS   141

18 EVOLUTION IN ACTION: MYXOMATOSIS   149

19 HERPES AND MORE ABOUT LATENCY   162

20 CONGENITAL VIRUS INFECTIONS   173

CONTENTS

21   VIRUSES AND CANCER: FOWL TUMOURS   180

22   VIRUS-TUMOURS IN MAMMALS – AND FROGS   186

23   NOT REALLY VIRUSES   195

24   MAN AND VIRUS ECOLOGY   202

Appendix 1   207

Appendix 2   208

Glossary   211

Bibliography   214

References   216

Indexes   227

Plates

[*between pages 120 and 121*]

1 Sections of tissue-cultures of rabbit testis showing changes produced by a virus
2 Electron-micrograph of virus particle of Newcastle disease of fowls
3 Helices released from the interior of measles virus
4 Electron-micrograph showing particles of fowl-plague virus
5 Three adenovirus particles showing surface structure and projections at each corner
6 Numerous round virus particles from a human wart
7 A herpes virus particle showing hollow prisms on the surface
8 Plate covered with a confluent growth of human cancer (HeLa) cells
9 Plate showing 'plaques' produced by bacteriophage
10 Leaf of *Nicotiana glutinosa* manually inoculated with two forms of tobacco mosaic virus
11 Bacteriophage
12 Cell in culture infected with an adenovirus
13 Bacteriophage particles being liberated by the bursting of an infected bacterium
14 Influenza virus particles attached to a fowl red cell
15 Filamentous and bizarre forms of influenza B virus
16 Section of mouse skin infected with ectromelia
17 Cells infected with an adenovirus and stained with fluorescent antibody
18 Cells with many nuclei infected with syncytial virus
19 *Culicoides riethi*, female feeding
20 *Aedes aegypti* female, in the act of feeding
21 Tower in the Zika forest, Uganda
22 Heronries round a pig farm near Tokyo
23 Adult *Haemaphysalis* ticks waiting to attach themselves to a passing animal
24 Clusters of *Haemaphysalis spinigera* larvae on the underside of leaves
25 Electron-micrograph of vesicular stomatitis virus
26 Rod-shaped virus particles being liberated when a nuclear polyhedron from the gypsy moth is dissolved with weak alkali
27 Larvae of tortoiseshell butterfly dying from infection with a nuclear polyhedrosis
28 Tortoiseshell butterfly larva showing liquefaction as a result of polyhedrosis infection
29 Tobacco plant systemically infected with a yellow form of cucumber mosaic virus
30 Leaves of tomato systemically infected with the tomato aucuba strain of tobacco mosaic virus

31 Stylet of the aphid *Myzus persicae* penetrating the phloem tissue
32 Pods of cacao, normal and infected with cocoa swollen shoot virus
33 Shoot of Arran Victory potato grafted to tomato symptomlessly infected with potato paracrinkle virus
34 Cambridge Favourite strawberry plants, normal and infected with arabis mosaic virus
35 Adult female *Xiphinema diversicaudatum*
36 Zoosporangia of *Olpidium brassicae* in root cells of lettuce
37 Spread of scrub at Lullington after myxomatosis had depleted the rabbit population
38 A natural case of scrapie in a Swaledale sheep
39 The development of warts after rubbing virus into the skin of a rabbit
40 A rabbit in which warty growths on the flank have developed into invasive cancer
41 Electron-micrograph of a thin section of a pellet containing many particles of a leukemia virus of mice

Acknowledgements

Plate 1, British Journal of Experimental Pathology; plates 2, 3, 4, 6 and 25, Professor A. P. Waterson and Mrs J. Almeida; plates 5, 7 and 15, Dr R. C. Valentine; plates 8, 14, 39 and 40, National Institute for Medical Research; plates 9 and 13, Dr D. E. Bradley; plate 10, Rothamsted Experimental Station; plates 12, 16, 17, 18 and cover photograph, Dr J. A. Armstrong; plates 19 and 20, S. A. Smith; plate 21, Royal Entomological Society; plate 22, Dr W. E. Scherer; plates 23 and 24, Dr T. R. Rao; plates 26, 27 and 28, Virus Research Unit, Agricultural Research Council, Cambridge; plates 29-36, D. B. D. Harrison and Rothamsted Experimental Station; plate 37, Dr A. S. Thomas; plate 38, Institute for Research on Animal Diseases, Compton; plate 41, Dr E. de Harver

The diagrams for this volume have been prepared by Design Practitioners from the following sources: Figs 1, 8 and 9, *Lancet*; Fig 2, *American Journal of Hygiene*; Fig 3, Royal Society of Medicine; Figs 5 and 6, World Health Organization; Fig 7, *Practitioner*; Fig 10, *Perspectives in Virology*; Fig 12, Proceedings of 2nd International Polio Congress; Fig 14, Dr C. E. Gordon Smith; Figs 15 and 16, *Royal Institute of Public Health Journal*; Fig 17, Dr E. S. Tierkel; Figs 18 and 19, Ministry of Agriculture, Fisheries and Food; Fig 20, *British Medical Bulletin*; Fig 21, *Journal of Hygiene*; Fig 22, Dr Miles Williams.
The author would like to thank Dr B. D. Harrison for much help and advice concerning chapters 15 and 16.

Part 1
Viruses in General

Chapter 1

Introductory

VIRUSES HAVE BECOME the subjects of a very important and fairly new branch of biological study – virology. The subject attracts people of very different interests. Many people are fascinated by the remarkable position which viruses hold 'on the borderline between life and death'. The chemical composition of the simplest viruses is beginning to be understood and possibilities are being unfolded of modifying them by chemical manipulations. Many of them, as we shall see, are of very simple structure, corresponding to three-dimensional geometrical figures, readily packed together to form crystals. No wonder, then, that people have been reluctant to think of them as living organisms. An important feature of viruses is their inability to multiply except within cells of higher beings, their increase being intimately connected with the life-processes of those cells. Study of the way they multiply or are multiplied has actually thrown light on the processes of multiplication of the living cells of animals and plants. Virology, therefore, is affording us insight into really fundamental mechanisms in biology and is indeed one of the most actively growing points of biological knowledge today. Study of these things and speculations as to future advances have resulted in the presentation to the public of very many books, articles in magazines, radio and television programmes concerning this exciting and rather glamorous side of science.

There has been corresponding neglect of an equally important and fascinating side of virology – its natural history and, accordingly, this is the aspect of virology to be dealt with in this book. Viruses may be thought of by some people as living crystals, but in practice they behave like parasites, having much in common with bacteria and other small parasitic organisms. The losses they inflict run into millions of pounds. Not only do they cause serious, often fatal or crippling, diseases in man but they are equally harmful to his herds and flocks and to his cultivated crops. The development of the sulphonamide drugs and antibiotics has given us the means of dealing with many bacterial infections; so the virus-diseases which are not yet controllable by these remedies loom relatively larger as causes of misery and economic loss. It may be that we shall overcome the menace of hostile viruses by methods arising out of the fundamental studies described in the preceding paragraph. Yet it is equally likely, perhaps in the short run more likely, that good results will be based on better knowledge of virus natural history. This is a study, moreover, of much intrinsic interest in itself. We shall have to look in the following chapters at the occurrence of colds in polar explorers, at the biting habits of mosquitoes in jungle tree-tops, at caves full of quarrelling bats, at nestling herons in Japan and at cancers in white mice. In this first chapter I shall try to make plain some of the many aspects of the subject and the underlying principles: in later chapters these will be taken up in greater detail and illustrated by telling the story of particular virus infections.

Reference will be made in chapter 2 to diverse views on the origins of viruses. We will for the present assume that they have arisen and that they have to make their way in the world, their evolution being determined by the process of natural selection. Here viruses start with an enormous advantage; they can multiply so rapidly and produce such millions of descendants within a short time that the forces of natural selection should be able to play upon them and permit their evolution at a speed leaving large organisms right back at the starting-post. Some kinds of bacteriophage, viruses which are parasites of bacteria, can, within the space of twenty minutes, destroy the cell they are infecting, liberating some hundreds of descendant virus-particles. The evolution of viruses can in fact be seen happening almost before our very eyes. In the thirty years since their discovery, influenza viruses have undergone demonstrable changes not once but many times.

Viruses, like other parasites, have to find ways of entering a susceptible host or host-cell, of multiplying within that cell, of getting out again and surviving in the outside world until they can get back again into another host. We shall consider briefly what happens to viruses inside an infected cell, but virus natural history is more particularly concerned with the means by which viruses get out, get around and get in again. There are five chief ways in which viruses are transmitted. We have, first, the respiratory route: virus gets out perhaps as a result of a sneeze and is carried through the air to the next victim. Or again, the virus is taken in with the food and passes out via the faeces. Infection may be mechanical, the skin or superficial tissues of an animal or plant being the portal of entry. A very important route is through the agency of a living carrier, or vector, of infection. Most vectors of virus-diseases are arthropods: in the case of animal-pathogens these are most frequently mosquitoes or ticks; while aphids, leaf-hoppers and other insects carry plant-viruses. Other vectors may be concerned, such as dogs and other biting mammals in the case of rabies and parasitic plants of the dodder family and even nematode worms for some viruses affecting plants. Finally, there is what has been termed vertical transmission; instead of infecting the neighbours of an infected host, infection is transmitted from parent to offspring either through the milk or even in, or in very close association with, the germ-cells themselves. The chapters in part 2 of this book will be looking in more detail at variations on these five themes.

To learn how to get out and around and in are the first things a parasite must learn, but those are only elementary lessons; it must learn also not to kill the goose that lays the golden eggs. Any parasite, or indeed any predator for that matter, which is too successful in attacking and destroying its victim will leave nothing for the next generation. So it comes about in the course of evolution that successful parasites are those which do not greatly harm their hosts; there develops a state of mutual toleration between parasite and host. The equilibrium which develops may be such that no overt disease results and the existence of parasitism would be undetected were it not for the prying research-worker. On the other hand a certain amount of disease may be desirable from the virus's point of view, to aid in its spread. Thus the madness of a dog infected with rabies causes it to rush around and bite and so infect other animals. The effectiveness of this is ensured by presence of rabies virus in the saliva. Again, respiratory virus infections lead to sneezing and, as a Ministry of Health poster told us during the war, 'Coughs and sneezes spread diseases'. Beet plants infected with yellows virus become as a result of infection more attractive to aphids; there are thus more of these available to spread the infection. Rabbit myxomatosis affords an excellent example. A too rapidly fatal infection gives less opportunity for the mosquito-biting which, at least in Australia, is the most important agency of spread. On the other hand, too mild a disease leads to there being less virus for mosquitoes to pick up. In consequence evolutionary pressures tend to stabilize myxomatosis at a particular level of virulence, not too high and not too low. Incidentally here is another example of the speed of evolutionary change which is possible. Chapter 18 will deal with this fascinating subject more fully. These examples are, however, exceptions to the general rule. Balance between a parasitic virus and its normal host is the commonest state of affairs. It will be clear that natural selection will always tend to bring this about. The host obviously gains by not being killed or damaged by a parasite and so selective forces will favour the emergence of a more resistant race. The virus gains by not killing off too many of its victims thus leaving plenty for succeeding generations. But that is not all: even a relatively harmless virus must not be too greedy and ambitious. The survivors of

5

a virus attack will commonly be immune to further infection. So a too-successful virus-epidemic is liable to leave no source of food for the descendants of the viruses concerned, just as surely if the infection is a mild one as if it is lethal. A dead host and an immune host are equally useless from a virus's point of view. It is one of the more fascinating aspects of the study of virus evolution to see how viruses have had to evolve in the direction of not being too efficient. We shall come across numerous examples of what we may call the benefits of moderation.

Our bodies, particularly our mouths, throats and intestines contain far more harmless bacteria than pathogenic, disease-producing, ones. Only within the last few years has it become apparent that in the virus-field also virus-infection is far commoner than virus-disease. There may be, as is likely with many enteroviruses, a transient symptomless infection of the gut especially in childhood; or a latent virus-infection may be more or less permanent. The situation is not quite the same with the viruses as with bacteria, as many of the latter are mere scavengers, subsisting on dead discarded cells, food-scraps and so forth, while no virus would so demean itself; all are intra-cellular parasites, even if harmless ones. How the state of latency is brought about will be discussed in later chapters.

If there is always, or almost always, a tendency to mutual toleration between virus and host, why don't we just have harmless virus-infections and not virus-disease? The answer is that there is never complete peace between the parties concerned, but only an armed truce. As in the world at large, something may happen at any time to precipitate an open conflict. In a world where evolutionary change is always going on it is too much to expect that every local balance of power will be stable for ever.

One form of disturbance is the introduction of a new species into an area. Normally, the native species in an area are not subject to disease as a result of infection by the local viruses. This may be for one of two reasons. Most animal viruses have evolved in association with certain species or families of vertebrates. Many diseases of birds do not infect mammals and most of those affecting man and monkeys do not naturally attack rodents or ungulates, though the virologist may train them in the laboratory to infect strange hosts. Some viruses affect only single species in nature, though this is unusual. So, many animals have nothing to fear from most of the viruses around them. If they ever were susceptible – and here is our second reason – they will have usually become resistant as a result of the evolutionary trend towards tolerance already discussed. It is when a species foreign to the environment comes along that the hitherto unobserved viruses infect him and produce dramatic and troublesome illnesses.

Such infections in introduced species are most commonly those due to the viruses which are transmitted by mosquitoes or ticks, the arboviruses to be discussed in chapters 8 to 13. The biting vectors concerned are often not particular as to what vertebrate they bite and so readily transfer a virus from one species to another. As we shall see, some viruses in North America are harmless parasites of birds but may get carried over to horses and people, causing fatal encephalitis. The diseases which are transmitted to man from other species are often referred to as zoonoses. Intrusion of a new susceptible host into a trouble-free environment is very frequently due to human intervention. Man is always looking for fresh worlds to conquer, and when he breaks virgin soil anywhere, he exposes himself to the local viruses and other parasites. Worse than that, he often brings with him his flocks and herds or his camp-followers and other man-associated species. It is note-worthy that while native North American birds are but little troubled by the encephalitis viruses just referred to, fatal disease may occur in the intro-duced pheasant and in the house-sparrow, which in America is called the English sparrow. In Africa, introduced pigs contract a form of swine fever from the native wart-hogs [48], while cattle catch malignant catarrh through contact with gnu, otherwise called wildebeest [129]. Other in-stances will be referred to in later chapters.

It need not be the introduction of an entirely strange species which lights up a virus disease; a race of a native species coming from another part of the world may be susceptible. Introduced European breeds of cattle or pigs may go down in Africa with infections to which the African breeds are re-sistant. West Africa used to be called the white man's grave because of the frequency there of fatal yellow fever; yet the native Africans, as we shall see, are relatively immune.

Another event causing outbreaks of virus disease is the introduction of virus into an area where it is unfamiliar. Many years ago yellow fever was imported into North America and by boat into Europe and caused epi-demics there. More recently, in 1962, it spread overland to Ethiopia and severely affected communities which were not immune as a result of pre-vious contact [142]. Measles is troublesome enough in its own territory but when introduced into virgin populations such as the Faeroes and Fiji, it affected whole populations at once and caused many deaths [32].

Man is responsible for unwittingly encouraging virus diseases in another way. In most natural communities there is a very mixed population of dif-ferent species of mammals, birds, insects and plants. Man has a passion for creating what a bacteriologist would call a 'pure culture' of a particular species. He keeps large herds of cattle or of sheep, he plants whole fields

7

with nothing but potatoes or beet and he himself crowds into towns. A wild animal or plant when infected with a virus is to some extent insulated from susceptible members of its own species by all the other kinds of animals or plants around it. In the relatively 'pure cultures' which man creates there is every opportunity for rapid passage of infection from one individual to another. Such rapid passage is just the right prescription for favouring an infecting virus and exalting its virulence. When the species crowded together are young and non-immune, the danger is particularly great. For such reasons children coming together in boarding-schools in the winter term have a very good chance of catching infectious diseases; and this applies also to newly enlisted service recruits.

There are yet other ways in which an upset in the balance of nature may be followed by virus diseases. When jungles are cleared for planting crops or building camps or towns, the relative numbers of different species are materially altered and abnormal spread of infectious agents may lead to disease. Seasons of abnormal weather may favour certain species such as mosquitoes, which may spread infection to new areas.

These are all things liable to cause outbreaks of infection, factors affecting whole communities. There are also factors operating in individuals. Many virus-infections are normally quite inapparent, the subject acquiring the infection for a brief period and becoming immune without any knowledge of being infected at all. Or he may have a mild fever and be just off colour for a day or two. A good example is afforded by poliomyelitis: in a small number of unlucky people, instead of causing trifling symptoms or none at all, the virus escapes from its normal habitat in the intestinal canal and reaches the central nervous system, there to cause paralysis. In the case of infection with the Japanese encephalitis virus, it is believed that only one out of hundreds or even thousands of infections of Japanese children actually leads to damage of the central nervous system. We do not know why a few are unlucky. They may be constituted just a little differently in their inherited make-up; or an unfortunate accident, perhaps a concurrent infection with another agent, may render them susceptible. There is evidence that louping-ill virus which affects sheep, particularly in Scotland, causes paralysis particularly when there is simultaneous infection with another agent, that of tick-borne fever [65]. Again, a chronic latent infection may be activated by certain stresses so that symptoms develop. The classical example, to be described in chapter 19, is that of herpes simplex or fever blisters. This virus, usually following a childhood infection, lies latent in the tissues of the face and is activated by various stimuli, often a fever or a cold; it then appears in the form of a crop of blisters around the nostrils or lips. There are many examples of the same sort of thing. Dr Richard Shope has described how pigs in Iowa may be latently infected with the virus of swine-influenza [147]; a sudden change in the weather, with wet and cold, activates the virus so that coughing and sneezing start simultaneously among the pigs in farms throughout a wide area. How far something like this affects our own colds and flu is a difficult subject to be dealt with in chapter 5. A controversial matter is whether viruses play a part in causing cancer in human beings, as they certainly do in chickens and in mice. In so far as they do it will certainly prove to be a matter of activating a latent virus-infection (see chapters 21 to 22).

A special instance must be mentioned, in which a virus-disease seems to appear from nowhere. Until the development of methods of growing viruses in fertile hens' eggs and in tissue-cultures, most studies of viruses affecting animals involved injection of infected material into rabbits, mice or other species, and observing whether the disease was reproduced or other changes caused. When this is done, it often happens that nothing obvious occurs at first. If, however, tissues from the injected animal are ground-up and inoculated into a second animal, obvious effects may be seen; or it may take several of what are called 'blind passages' or 'passages by faith', before the injected virus is sufficiently exalted in virulence or adapted to a strange host for anything to be apparent. The technique has often proved most valuable, especially in adapting all sorts of viruses to grow in mice. There is, however, a snag. Many, perhaps most, animals are apt to be latently infected with their own viruses; a dozen or more are known to occur in mice. The technique of blind-passage is only too likely to bring to light one of these latent infections. Many workers in the course of research on viruses have produced illness in their inoculated animals after a few blind passages, and this has ultimately proved to be due to some other virus than that with which they started. Only too often this has led them astray and caused them to make unjustified claims to have isolated an important virus.

I was myself concerned in what was a very early, perhaps the first, trouble of this kind. More than forty years ago rabbits rather than mice were the popular animal for virus research; and Noguchi, working at the Rockefeller Institute, had found that inoculation into the testes of male rabbits was a useful method of study. When I went to work at that institute in 1923, Drs Rivers and Tillett [134] had been using this method in their search for the virus of varicella or chickenpox. After a few blind passes inflammation was produced in the rabbits' testes, and microscopical study of these showed changes in the nuclei of cells which were quite dramatically like those seen in chickenpox (plate 1). The effects, however, were not neutralized by convalescent serum from chickenpox patients; so Rivers and Tillett were, very fortunately, cautious in drawing conclusions.

Drs Homer Swift, Phillip Miller and I were working on rheumatic fever in the same building. Similar blind passage experiments led to similar results, this time when we began with rheumatic fever instead of chickenpox material. We had also injected our supposed rheumatic fever virus into the chest-wall of rabbits and produced an inflammation of the pericardial sac around the heart, not unlike that seen in rheumatic fever. For a short while our hopes were high. We soon became suspicious, however, on discovering that the changes in the rheumatic fever rabbits' tissues were identical with those in the chickenpox rabbits. It turned out, of course, that we were activating a virus latent in the rabbits, one which is not known ever to do rabbits any harm in the ordinary course of events [11].

This sort of complication is particularly bothersome for workers in cancer research, where tumours of mice have yielded much important information. For latent viruses seem to flourish particularly in cancerous tissues. Some viruses of mice are capable of producing in them leukaemia, which is essentially a cancer of blood-forming tissues. On a number of occasions, such leukaemia-viruses have been found growing in other sorts of cancers, with the causation of which they have nothing whatever to do; no wonder that the poor cancer-worker has been thrown into much confusion.

Virus-disease revealed by research workers in artificial ways such as these comes rather doubtfully into the category of natural history, unless of course one regards the virologist as an ecological factor to be considered along with the weather and the other matters we have dealt with. We have in any case had to consider the matter in order to round off this first look at the subject.

I have attempted in this chapter to take a preliminary bird's-eye view of virus natural history, or, if you prefer, virus-ecology, the study of viruses in relation to their environment. Before considering individual viruses and particular aspects of the subject in more detail, it will be necessary to go back a little to consider what viruses are, of what and how they are constructed, how they multiply within a cell, how they affect such a cell and what are the mechanisms of defence against them. The reader can then better understand the matters to be described in the truly natural history chapters from chapter 5 onwards. Readers who are already familiar with the elements of virology may prefer to skip the next three chapters.

Chapter 2

The Nature and Variety of Viruses

AS WE SHALL SEE in a moment, there are divergent views as to the essential nature and origins of viruses. There is much more knowledge about their structure and chemical composition.

All viruses contain nucleic acids and proteins; the larger ones are more complicated. Both these kinds of organic substances consist of long chains of molecules. The fundamental components of the nucleic acids are pyrimidine and purine bases combined with sugars which may be ribose (ribose nucleic acid = RNA) or de-oxyribose (DNA). I will not go deeply into the chemistry, as this is not that kind of book. Animals and plants, including bacteria, all contain both kinds of nucleic acids; viruses are peculiar in that, so far as we know, they only contain one kind. Plant viruses all contain RNA, bacteriophages with a few exceptions contain DNA, of insect viruses some contain one, some the other and the same applies to the viruses of vertebrates. The long chains of DNA are usually double-stranded while RNA is generally single-stranded. There are, however, some animal viruses (reoviruses), and some structually similar plant-viruses which contain double-stranded RNA [64]. Neither RNA nor DNA must be thought of as single chemical compounds; they are families of compounds and variations in the make-up and arrangement of the constituent bases determine the differences between one virus and another. On the nucleic acids depends the hereditary transfer of characters from one generation to the next; as viruses change and evolve there must be changes in their nucleic acids.

The nucleic acids are rather labile substances, readily inactivated by certain enzymes. They can be protected from the action of these enzymes by proteins wrapped around them. This is one of the main functions of at least some of the proteins in a virus's composition. With some viruses it has been possible to infect a cell and induce virus-multiplication with the naked nucleic acid, but one has to be very slick in achieving this before the nucleic acid can get knocked out by a hostile enzyme. Another function of the protein is to make the specific contact with subtances on the surface of the host-cell which permit the virus's ready entry. It is also this protein which stimulates the body to make specific antibodies against it. The protective protein surrounds the virus in one or other of two ways. There may be a regular box or cage round the virus and this is called a capsid; it is made up of a number of similar units called capsomeres, arranged in a regular geometrical pattern with the nucleic acid coiled up inside the box [183]. There is probably a close structural relation between the components of this coil and the components or capsomeres of the box. One property used in classifying viruses concerns the numbers of capsomeres in the capsid. In an alternative method the protein is arranged round the nucleic acid in a coil or helix, so that the essential part of the virus is a long thin structure. The virus may thus turn out to be a rod, as are many plant-viruses. Or the long structure may be coiled up within an outer shell (plates 2 and 3); there may be a number of similar units in this shell, as in the myxoviruses, but they are not in a regular geometrical arrangement as are the capsomeres of other viruses. The largest true viruses, those of the poxvirus family, are structurally more complex and may not fit exactly into either of these two categories.

We have now been introduced to two characters concerned in classifying viruses; the nature of their nucleic acid, whether RNA or DNA and the relative arrangement of the protein and nucleic acid in their nucleoproteins, as a box or as a coil. A third character concerns the presence or absence of an outer envelope beyond the nucleoprotein core. Some viruses (picornaviruses) and many plant viruses seem to consist wholly of nucleoprotein, while others have an outer envelope often containing lipid (fatty) material. Such enveloped viruses are usually inactivated by treatment with a bile salt (desoxycholic acid) or with fat solvents such as ether and chloroform, and here we have a very simple test which helps in putting viruses into one or another group. Some viruses, as they leave the cell they have infected, actually incorporate into their outer membranes some substances derived from their host-cell, but this does not imply any hybridization with their host cell, any more than we hybridize with sheep when we put on a woollen overcoat.

The use of these few fundamental characters in classifying viruses is something rather new. Until a decade or so ago the custom was to use quite different properties – the species or plant of animal attacked, the pathological changes produced, the symptoms caused. These things are still used in some textbooks for convenience – viruses causing diseases of the central nervous system, of the skin and so on; it is now realized, however, that these properties are of little value for classification: a virus can change its properties, naturally or as a result of manipulation in the laboratory, so that it infects different species or different systems. Yellow fever, which naturally damages the liver of man and monkey, can be turned into something affecting the brains of mice, and similar changes can be induced in many other viruses; but no one has yet succeeded in modifying the more fundamental characters on which emphasis is now placed. There is a surprising consequence of the new importance given to virus-structure. Until recently those studying viruses of animals, plants and bacteria have belonged to essentially different disciplines. But if one no longer considers affinity for mouse-brain or human liver as valid criteria for subdividing and classifying viruses, is it reasonable to continue to separate the viruses of the cat and the cucumber? In fact it turns out that some animal and plant viruses are indistinguishable so far as size and shape are concerned. It can of course be argued that this is a coincidence; there may be a limited number of ways in which biologically active materials such as nucleic acids and proteins can be put together; the resemblance might simply be due to what is called convergent evolution. This means that the forces of natural selection have impelled creatures of very different origins to come to resemble each other, often because they have acquired similar habits. One cannot, however, escape the fact that some viruses can multiply both in plants and in insects and others in both insects and vertebrates. So it would be very difficult to know where to draw the line if plant and animal viruses were considered quite separately.

Classification is generally considered a dull subject, though those who make it their business are apt to become passionate on the subject and to quarrel fiercely with other 'taxonomists'. Knowledge of relationships does, however, often come in very useful and enable one to predict that certain properties will be found in another species. Lately, many unexpected relationships have come to light between viruses of man and domestic animals. Close relations have been revealed in the following pairs or groups of viruses: in measles, dog-distemper and cattle plague; adenoviruses causing sore throats in man and a virus causing hepatitis in dogs; Newcastle disease of fowls and mumps; herpes in man and pseudo-rabies or mad itch of cattle; Coxsackieviruses affecting man and the virus of foot-and-mouth disease.

It will be well for the reader to have some mental picture of the chief virus families, for they will be very frequently referred to throughout the book. Appendix 1 will give the outlines of a proposal for classification of all viruses – those affecting vertebrates, insects, plants and bacteria [130]. Not all viruses are represented in the scheme, for there are many about which information is not yet adequate. The proposal was put forward by an international committee, but is in no way official: with more information coming along it is certain to be modified later. Appendix 2 lists the more important animal viruses arranged so far as possible in their families. Bi-nominal names, of genus and species, as used in the Linnaean system for animals and plants, are not in general use for viruses as yet; but they may very well come into use before long.

Before we look quickly through the virus-families another word is necessary about these capsomeres. Much has been written about the way they are arranged; rather surprisingly solid geometry comes into it [81]. Many viruses, formerly thought to be spherical, are now seen to be icosahedra, figures with twenty triangular faces. Capsomeres are arranged 2, 3, 4, 5 or more along each side of each triangle. These would fit best together if each were hexagonal in form. However, a lot of hexagons fitted together would make a flat sheet and wouldn't conform to the surface of a solid structure. This is achieved, however, if the capsomeres placed at the various vertices are pentagons. There is evidence, in the case of an adenovirus, that the capsomeres at the vertices are of different chemical composition from the rest and exist as projections (plate 5). For most viruses of this sort the numbers of capsomeres on a virus surface can be calculated from the formula $10(n-1)^2+2$, where n is the number of particles on the sides of the triangles. Where $n=1, 2, 3, 4, 5, 6$ and 10 we find capsomere numbers of 12, 42, 92, 162, 252 and 812. All these numbers are in fact found in actual viruses, 12 for certain bacteriophages, 42 for wart viruses, 92 for reovirus, 162 for herpesvirus, 252 for adenovirus and 812 for a virus attacking crane-flies (*Tipula* or daddy-long-legs).

Virus manufacture within a cell doesn't always go according to plan. Some hollow capsids are apt to be formed lacking the core of nucleic acid – presumably the cell ran out of nucleic acids while still having enough protein components to go on making capsids. There may also be formed in some virus-infections tubular structures instead of properly formed capsids; it is suggested that this happens when the cell tries to make virus particles entirely from hexagonal compounds, the pentagons being in short supply. It remains to add that the capsomeres themselves seem to be composed of still smaller subunits. Now to our review of the families of viruses infecting vertebrates and first for those composed of ribose-nucleoprotein (riboviruses).

The *Picornaviruses* embrace the smallest known viruses including those of poliomyelitis and foot-and-mouth disease. The name is derived from 'pico' a prefix meaning very small plus the initials RNA to remind us that they are RNA-containing viruses. It is likely that there will turn out to be some equally small DNA-viruses (deoxyviruses), but they will not concern us in this book; the name *Parvovirus* has been proposed for them. The picornaviruses are roughly spherical viruses, about 30 mμ in diameter. 1 mμ = 1/1,000 μ or 1,000,000 of a millimetre, so picornaviruses come out at somewhere near a million to the inch. They seem to have a regular arrangement of capsomeres on their surfaces but there is still some doubt as to how many there are; it might be 32, 42 or 60. They may not be arranged according to the plan of symmetry we have been considering. There are no outer membranes, so picornaviruses are not put out of action by fat solvents. The relevance of these and other characteristics along with those of other animal virus families will be clear from Appendix 1.

The picornaviruses which have been best studied are the enteroviruses or viruses of the intestines; this is a term which includes the viruses of poliomyelitis, the echoviruses and Coxsackieviruses, all of them inhabitants of human alimentary canals and many of them liable at times to invade and damage the central nervous system. Knowledge about all these viruses took a great step forward when J. F. Enders and his colleagues [55] in 1949 discovered that poliomyelitis would grow in culture in all sorts of human and monkey tissues. In the great surge of interest in the practical aspects of tissue-culture, to be discussed more fully in chapter 3, the echoviruses amongst others were first brought to light. They were named from the initials E, C, H, O, standing for enteric, human, cytopathogenic (i.e. cell-damaging) and orphan. The term orphan was applied to viruses which could be recognized in the laboratory but were not associated with known disease. This application soon ceased to be appropriate, for a number of echoviruses were incriminated as causing fevers with rashes or with meningitis. The Coxsackieviruses are so called from the village in New York State where they were first isolated. They can cause a kind of sore throat (herpangina) or a fever with pains in the chest or abdomen – (pleurodynia or Bornholm disease). Their striking characteristic from the laboratory point of view is that they produce fatal illnesses when injected into newborn or suckling mice. Evidently delighted with having invented the word ECHO-virus, virologists proceeded to construct the names ECBO, ECSO, ECMO for enteric cytopathogenic bovine, swine and monkey 'orphan' viruses. These names are unlikely to endure, for picornaviruses are being recovered from the intestines of many different animal species and the twenty-six letters of the alphabet will soon be exhausted. It will probably be better to talk of bovine etc. enteroviruses. Their naming and classification will not be easy, for there is no certainty that the viruses affecting one species are all different from those affecting others; indeed one human echovirus has been recovered from dogs. Most of these viruses isolated from intestines seem to be pretty harmless, but there are some which are less so. One of those from

15

swine causes a paralytic disease something like polio (Teschen disease), while one of those of chicks is responsible for a serious disease, epidemic tremor.

Poliomyelitis viruses are, as is generally known, of three types which can be separated according to their antigenic characters by serological tests, as described in chapter 4. Similar tests sort out the echoviruses into thirty-three types and the Coxsackies into at least thirty. There are many types also of those affecting species other than man. Another picornavirus of importance is that of foot-and-mouth disease with seven main types (see chapter 14). Chapter 5 will deal rather fully with the rhinoviruses causing colds; these are picornaviruses adapted to growth in the nose rather than the gut. They have only been recognized as a group since 1960 and are of very many serological types. Besides the human pathogens there are rhinoviruses infecting calves, horses, cats and doubtless other species also.

The *Reoviruses* constitute a small family of RNA-containing viruses with 92 capsomeres. They occur in animals of many species but their role in causing disease is obscure. As already mentioned they, in common with certain plant-viruses, have an unusual two-stranded RNA in their composition [64].

Arboviruses, also referred to as arborviruses, derive their name from a telescoping of 'arthropod-borne viruses'. The second R in arborvirus was dropped, lest people should think the name had something to do with trees. The name arbovirus has been used for those viruses with a biological cycle of development in both arthropod and vertebrate and has thus not been based on the fundamental criteria used for defining the other families. It seems almost certain, however, that the vast majority of included viruses will turn out to belong to a natural group of RNA viruses with outer membranes – and therefore ether-sensitive. The symmetry of the nucleoproteins of these viruses, whether cubical or helical, is unsettled. They vary in size from something not much bigger than picornaviruses, 30mμ, to something with a diameter of about 100mμ. They are divided on an antigenic basis into groups – A, B, C and several others. Besides four major groups, there are smaller ones, groups of three, two and singletons. Many are still unclassified. Arboviruses thus form the largest known family of viruses of vertebrates with at least 170 known members and more being discovered every year. There will be a good deal about arboviruses in this book, the A group in chapter 8, the B's in chapters 9 to 12, for there are fascinating aspects of the interactions of virus, vertebrate host, arthropod vector and other agencies which disturb the environment. In fact, the natural history of the arboviruses could fill a book by itself.

Next come the *Myxoviruses* with a helical symmetry for their nucleo-proteins. They are larger than the viruses so far considered but as they are surrounded by an easily deformed membrane one cannot give an accurate figure for their size; it varies between 60 and 300 mμ (plates 4 and 15). This membrane contains lipids, so the myxoviruses are sensitive to ether; moreover it commonly contains components coming from the host cell. All over the virus surface are radiating spikes, but not regularly arranged like the capsomeres of a capsid. These spikes are made of a protein and, for most of the myxoviruses, are concerned with making contact with and entry into a susceptible cell. This property is reflected in the fact that they also have affinity for the red blood cells of some species (plate 14). The virus particles therefore adhere to these and, by forming bridges between one red cell and another, cause the red cells to clump together as cells are clumped or agglutinated in blood-grouping tests [78]. The phenomenon is called haemagglutination and the spikes represent the haemagglutinin. They carry an enzyme which digests the mucin or muco-protein on the surface of red cells. This affinity for mucin gives the myxoviruses their name, for the root myxo- denotes mucin in Greek. The haemagglutination was the most characteristic feature of the myxoviruses first described – influenza viruses A, B and C, mumps and Newcastle disease of fowls. Subsequently other viruses were found, resembling typical myxoviruses morphologically but lacking the power to clump any of the red cells yet tested. Three related viruses coming to be classified as myxoviruses are measles, dog distemper and rinderpest or cattle-plague; haemagglutination has only been shown for the first of these. Some viruses, known as para-influenza viruses 1, 2, 3 and 4, cause influenza-like illnesses in children or common colds. When first isolated these do not clump red cells readily but they can be grown in tissue-cultures. When red cells, usually those of guinea-pigs, are added to the cultures they stick to the virus which is being released from the surface of infected cells. They thus pick out those cells which are infected and form a pattern in the culture which can be revealed with low powers of the microscope; the phenomenon is called haemad-sorption [177]. An important myxovirus, which shows neither haemag-glutination nor haemadsorption, is the respiratory syncytial virus (plate 18), considered later as one of the agents causing common colds.

We now come to what we may call a mixed bag of riboviruses. They are all of about the size of myxoviruses, have external membranes and are accordingly ether-sensitive. It is suspected but not certain that their symmetry is helical, but it is harder to break open the core and see how it is constructed than in the case of myxoviruses. Some have spiky surfaces, but mostly they do not haemagglutinate. They will probably come to be classified in one or more groups, very close to the myxoviruses. The viruses concerned include those causing tumours and leukaemia in fowls; also leukaemia and some other diseases in mice and other rodents. Rabies too must be mentioned here; this virus certainly has a helical core, a little different from those of typical myxoviruses [3].

Four families of DNA-containing viruses (deoxyviruses) have to be considered – the adenoviruses, papovaviruses, herpesviruses and poxviruses. The *Adenovirus* group was only discovered in 1955. Drs Rowe, Huebner and their colleagues [139] at the National Institutes of Health in Bethesda near Washington, in cultivating cells from tonsils and adenoids of children, grew out a new kind of virus which destroyed the cultivated cells. Soon after, a similar virus was found causing outbreaks of sore throat amongst military recruits. The prefix adeno- given to the virus indicates an affinity for adenoid or glandular tissues. The presence of regularly disposed capsomeres – 252 of them – was first demonstrated in this group. Adenoviruses are about 70 mμ in diameter. There is no outer membrane and so these viruses are ether-stable. They develop within nuclei of infected cells, where they may be seen closely packed together in a sort of microscopic crystalline arrangement (plates 11 and 17). There are at least thirty antigenic types amongst those infecting man, while others are found amongst mice, cattle, pigs, fowls and doubtless other species. Reference has already been made to one causing jaundice and other illnesses in dogs, the virus of canine hepatitis. All the known adenoviruses contain a common antigen recognizable in serological tests – all, that is, except one from fowls. The adenoviruses often persist as latent infections in the throat, especially of children; but some of the types cause outbreaks of fever and sore throat particularly amongst recruits. Inflammation of the eyes – conjunctivitis – is sometimes associated with the sore throat, and one type, no. 8, is particularly troublesome in causing eye infection. A special incidence in industry has given the infection the name of 'shipyard eye'. A recent discovery, to be discussed in chapter 22, is that several types can give rise to cancer in hamsters. This is important, as it should make it clear that there is no special category of tumour-viruses. Viruses of several families may be responsible and here we have an instance where a fairly harmless respiratory virus has revealed an unexpected propensity for tumour-production.

The *Papovaviruses*, next on our list, are on the other hand, almost all concerned in causing growths, either benign ones such as warts or malignant cancers; they have also been called Papillomaviruses. They have cubical symmetry with 42 capsomeres, are mostly around 45 mμ in diameter and are ether-stable with no outer membrane. Their development, as with adeno-viruses, is in the nucleus, where they may be seen packed in a similar regular manner (plate 6). There are members causing warts (papillomas) in man, rabbits, horses, cattle, dogs and other species. One is the cause of tumours of various sorts in mice. The tumours produced by this virus are of such different sorts that the name polyoma 'many-tumour' was given to it. Another papillomavirus came to light when people began to use monkey kidneys in a big way for research on polio; it is very apt to grow out from apparently normal monkey kidneys, producing vacuoles in affected cells particularly in those of African monkeys. So it was called vacuolating agent. A name of the whole group, Papovavirus, was proposed; it was made up of the initials of some of its chief members – PA for papilloma, PO for polyoma and VA for vacuolating agent. Objections to the name were raised and 'papillomavirus' is preferred by some as a group name. Like the adenoviruses the vacuolating agent can give rise to tumours in hamsters. There is only one of the papovaviruses so far known, the K virus of mice, which is not known to cause growths of any kind, either warts or cancers.

The *Herpesviruses* are of a size larger, mostly 120–180mμ across. Their 162 capsomeres are in the form of hollow prisms (plate 7); models constructed to show this have a rather fearsome appearance. There is a membrane outside the capsid and all members of the group are sensitive to ether. Development of herpesviruses begins in the nuclei of infected cells and the outer membranes are acquired as the newly growing virus passes outwards. Cells infected with herpesviruses commonly show characteristic intranuclear 'inclusion bodies' of a type to be described in chapter 3. Included amongst the herpesviruses are that of *Herpes simplex* in man and closely related viruses affecting monkeys and marmosets. The monkey virus, known as B virus, is much feared by those using monkeys in work with poliomyelitis or for other purposes. It is not uncommonly carried by rhesus and other monkeys and a bite from such a monkey will lead to an infection of the brain and spinal cord which is almost always fatal. Not very long ago a king of Greece died as a result of a bite from an infected pet monkey. We shall have occasion to mention a related virus, that of pseudo-rabies or mad itch, affecting cattle and several other species. Other viruses in this family cause acute respiratory infections in cattle (infectious bovine rhinotracheitis), in cats (feline rhinotracheitis), in birds (infectious laryngo-tracheitis) and abortion in mares (equine rhinopneumonitis). Viruses of a subfamily of herpesviruses cause great enlargement of cells, often in the salivary glands and so they are called cytomegaloviruses – big-cell viruses. They affect very many species; the human representative is somehow congenitally transmitted and gives rise to illnesses of various kinds in babies, often leading to serious mental retardation, if not to death (see p. 173).

Finally we have the *Poxviruses* causing smallpox (but not chickenpox, which is due to a herpesvirus), cowpox, sheep pox, goat pox, swine pox, horse pox and so on. Myxomatosis of rabbits is due to one of this group. The poxviruses are the largest and most complex of the true viruses; they are mostly rather brick-shaped with one diameter rather longer than the other; the virus of vaccinia is about $200 \times 250\,\mathrm{m}\mu$. How their nucleoproteins are arranged, whether in cubical or helical symmetry is uncertain. There is a coiled structure to be seen in some members but it seems to be in the outer part of the virus, not in the core as with myxoviruses. There are also hollow spikes on the virus periphery. Although there is an outer membrane, some of the poxviruses are sensitive to ether while others are not. Perhaps the integrity of the membrane is essential for the infectivity of some members but not for others. Unlike the other DNA-viruses we have considered, poxviruses seem to undergo their development wholly within the cytoplasm of infected cells (plate 16). There is a puzzle concerning the origin of the vaccinia virus used for vaccination against smallpox. It was formerly thought to be derived from smallpox (variola) virus by adaptation to calves or sheep. However, nobody can achieve that now if they take proper precautions against accidentally picking up pre-existing vaccinia virus from the environment; earlier workers slipped up in this way. Alternatively vaccinia may have been derived from cowpox of cattle; this virus certainly can lead to protection against smallpox, as Jenner showed. Yet others think vaccinia is a sort of primitive generalized poxvirus from which many of the others evolved.

As we have seen, a number of major groups of animal viruses have been delineated with the help of a limited number of criteria, all concerned with their chemical composition and morphology. For finer degrees of classification we have to use other characters which, though useful and often indispensable, have to be applied with caution, since viruses can change so readily and many characters are far from stable. Some of those which have proved useful are mentioned in the following list:

Sensitivity to inactivation by certain chemical compounds.

Sensitivity to inactivation by heat and by changes in pH.

Ability to grow in fertile hens' eggs. Some viruses grow on the outer chorio-allantoic membrane, some in the cells lining the allantoic cavity, being released into the allantoic fluid, others in the amnion and some in the yolk-sac.

Antigenic characters, to be discussed more fully in chapter 4.

Size.

Ability to multiply in arthropod vectors.

Multiplication in nucleus or cytoplasm.

Affinity for particular tissues or host-species.

Production of particular pathological changes and symptoms.

These last two characters now receive less emphasis than any others; formerly they were stressed almost to the exclusion of everything else.

Not all viruses can be classified in the groups we have been reviewing. There are a good many not yet sufficiently well understood to be safely placed in any of the existing families nor yet allotted a new family of their own. Nearly all those dealt with in this book, however, fall into one of our defined families. An exception is the virus of infectious hepatitis (jaundice) of man, about the nature of which we know very little (p. 167). Mention will also be made of the 'viruses' of psittacosis and trachoma. These belong to groups formerly included with the viruses but now considered to belong more properly, together with the rickettsiae of typhus, amongst the smallest, most highly parasitic bacteria. As psittacosis and trachoma are so commonly referred to as viruses and as they illustrate some points in the natural history of small organisms they will be considered in chapter 23. The reasons for excluding them from the viruses are that they have something in the way of a cell wall with a component, muramic acid, which is present amongst bacteria but lacking in viruses; their development cycle includes at one stage 'binary fission', that is division of one unit into two; their metabolism is so like that of bacteria that they can be suppressed by some antibiotics which are active against bacteria; they contain, as bacteria do and viruses do not, both kinds of nucleic acid, DNA as well as RNA.

The viruses pathogenic for insects [154] have not been finally classified according to their fundamental properties. They are at present arbitrarily divided into those which form inclusion-bodies and those which do not; the inclusion-bodies may be relatively large crystalline 'polyhedra' within which virus particles are embedded; or they may be very small crystals or granules. The polyhedra-forming viruses are subdivided into those in which viruses, which are rod-shaped, grow within nuclei and those with almost spherical particles multiplying in cytoplasm. Only a very few viruses are known in the group which forms no inclusions. Two in this group actually infect arthropods other than insects – red mites, which belong to the arachnids. The viruses included in the cytoplasmic polyhedroses, referred to just now as 'almost spherical' seem in fact to be icosahedra like so many of the vertebrate viruses. Unlike the nuclear viruses which contain DNA, some of them seem to be riboviruses. The largest of them, however, the *Tipula* iridescent virus, is a deoxyvirus; this is the one which boasts 812 capsomeres. The insect pathogenic viruses are of especial interest as the phenomena of latency amongst them show features of special interest.

Two views are current concerning the essential nature and the origin of viruses. One view was put forward by Dr R. G. Green [66] in the United States and independently in Britain by my former chief, the late Sir Patrick Laidlaw [96]; this is often called the Green-Laidlaw hypothesis. It points out that all creatures which adopt a parasitic mode of life tend to discard those structures and functions which are useless to them, for unnecessary paraphernalia of existence tend to become a mere embarrassment. Fleas, which are descended from flies, have lost the wings they no longer need. Tapeworms find they can absorb their food perfectly well through their exterior surfaces, so they dispense with the luxury of an alimentary canal. Viruses may well be derived from bacteria or other micro-organisms. (We call them 'micro-' organisms but they are large in comparison with viruses.) These, it is suggested, have achieved a simplified structure by doing without a cell-wall and ultimately even without the enzymes concerned in energy-production and in various synthetic mechanisms. They use instead the enzymes of the cell they are infecting, and, in Laidlaw's words, lead a borrowed life. Burnet [31, 32] has suggested one way in which this may have come about. A number of bacteria are known to lose their cell-walls as a result of certain stimuli, forming what are called protoplasts. First of all as a parasite evolved, its capsule might be weakened allowing free passage of metabolites between parasite and host-cell; later it could be dispensed with altogether, allowing host and parasite substance to intermingle: the efficiency of deviating the cell's armoury of enzymes to build virus-substance would thus be much increased.

All this could have happened along several separate evolutionary lines. Some viruses could have been derived from bacteria, others from small protozoa such as malaria parasites. I have always been impressed by the fact that arthropods, particularly insects, play such a prominent role in nearly all the major virus groups; many viruses can multiply both in vertebrates and in the mosquitoes or ticks which carry infection from one to another; very many plant-viruses are transmitted by insects and some multiply in both insect and plant; there are also the large families of viruses which are pathogenic for insects. I have accordingly advanced very tentatively the suggestion that all sorts of viruses may have arisen from the symbiotic organisms which are found in the tissues of very many insects and are not infrequently necessary for their existence [7]. Bawden [15] and Black [22] have put forward similar views concerning plant-viruses.

All this is perfectly possible but we cannot prove it by looking into the past nor observing its occurrence today. The same applies to an entirely different view of virus-origins, a view having just as many advocates as the other [179]. According to this notion, a particle containing nucleic acid, with presumably some protective protein, might be accidentally transferred from one cell into another, perhaps of a different species. One could perhaps most readily visualize this as happening with the aid of an arthropod vector, but one could also imagine the engulfment of inhaled or swallowed material by phagocytosis as being concerned. Having got into a strange cell, the foreign material would normally be got rid of. Just occasionally, however, it might find its new environment not too hostile and it might be able to multiply in the manner to be discussed in the next chapter, its descendants finding their way out of the cell and in due course initiating another cycle in fresh cells. In other words a piece of cell would have become autonomous and attained the status of an independent being. After this, since it contained nucleoprotein as its genetic substance, something subject to variation, the forces of natural selection would inevitably work upon it and doubtless lead to the development of an organism better adapted to its new role in the world.

These two very different views of virus origins both have many advocates amongst virologists; one may conclude that it is very difficult to decide between them and such a decision will not be made soon, if indeed it is ever made. It can be argued that one reason for favouring the Green-Laidlaw hypothesis, at least for viruses of vertebrates, is that so many of them fall rather neatly into families with related viruses infecting animals of many different kinds. A common origin for members of each virus family seems probable and that seems to imply a history going back into the very remote past. One thing nearly everyone is agreed about – that it is unlikely that viruses arose out of primordial ooze ahead of other kinds of life. Their intracellular parasitism is so much a part of them that their origin from something other than other living cells seems difficult to imagine.

Chapter 3

The Virus and its Host

VIRUS ECOLOGY, the study of viruses in relation to environment, can be studied at three levels: how the virus fares in the cell in which it multiplies, what happens to it in the whole host, plant or animal, which it infects, and finally its progress and its fate in the world at large. Although this book will deal mainly with this last, wide aspect of virus ecology, some understanding is necessary of the other two views of the matter.

We must have some idea of how viruses enter cells, multiply in them and get out to enter other cells; otherwise we cannot fully grasp what happens from the wider points of view of the whole host and the community of hosts. Recent increase in our knowledge has stemmed in recent years from studies of viruses growing in tissue-culture. Here it is possible to isolate cells and viruses from at least some disturbing influences.

Living cells from animals were first grown outside the body in 1907; in 1913 vaccinia virus was shown to survive for some weeks in such cultivated cells, and in 1925 its multiplication in tissue-cultures was definitely proved by Parker and Nye [122]. Soon after, Dr and Mrs Maitland [106] showed that this virus would grow in minced-up tissue suspended in a fluid medium – no elaborate methods being necessary. In the earlier years of tissue-culture, separate investigators, working in an academic atmosphere, found out quite a bit about the behaviour of viruses in culture. In 1949, however, Enders, Robbins and Weller [55], in Boston, made a discovery which altered the whole course of work in this field: they found that poliomyelitis virus would grow in all sorts of tissues of man and monkey, tissues other than the neurones which alone it was thought to infect; moreover it would produce destructive (cytopathic or cytopathogenic) changes in the cells, visible with low powers of the microscope. It was thus possible to study the poliovirus quantitatively by a simple method: previously this had to be done laboriously and at great expense by inoculating material into the brains of monkeys. The new advance had a number of consequences: it led to a method of making a vaccine against polio; it led to the discovery of the echoviruses, adenoviruses and others which produce changes in cultures, although not readily detected by work in experimental animals or fertile eggs; it was followed also by the regular use of tissue-cultures for diagnostic purposes in routine diagnostic laboratories. Virology was no longer the preserve of a few specialists.

The tissue-culture techniques used in virology are now quite varied. For many purposes cells growing on the walls of test-tubes do better when slowly rotated on a 'roller drum' so that fluid bathing them constantly washes them, bringing food and oxygen and removing waste products. Often, however, this rolling is unnecessary. Cells may be grown in single sheets on glass, often with an overlying thin layer of agar; virus added to such cultures produces small foci of cell-destruction visible to the naked eye and this offers a simple method of titrating virus quantitatively (plate 8). The method gives results similar to those obtained when dilutions of bacteriophage are 'seeded' on to bacterial cultures and counts made of the resulting clear 'plaques' (plate 9); or when dilutions of a plant virus are painted on to leaves of a suitable test plant (plate 10). Again, pieces of tissue in which cells remain in their natural relationships can be kept alive in culture for weeks at a time. Cells in these 'organ cultures' remain in something nearer to natural conditions than in ordinary tissue-cultures.

Embryonic tissues usually grow better and survive longer in cultures than those of adults, but adult tissues grow well enough for many purposes. Kidney cells do particularly well, giving good growth of epithelial cells and not getting too heavily overgrown with fibroblasts as many other tissues do; cells from monkey kidneys are therefore used in a big way in work on polio. Cells from some cancers have been got going as indefinitely growing cell-lines; the HeLa cells, named from the letters of the name of the cancer patient who provided the cells, have been used in laboratories all over the world for more than thirteen years. Sometimes tissues from a normal person or animal can be adapted to give indefinite growth in culture, but when this happens the cells are found to have abnormal chromosomes together with other characters associated with cancer. Ways have, however, been found lately of growing normal cells for many generations without appearance of chromosomal abnormalities.

Some unexpected things have turned up. Polio virus, as mentioned, grows well in a monkey's kidney cells, but inject it into the kidneys of a live monkey and it will not grow at all. Still more surprising, many viruses will grow, in tissue-culture, in cells of an animal species which is quite insusceptible in nature. Thus the virus of myxomatosis of rabbits will multiply in cultures of tissues of guinea pigs, rats and even man, though naturally it only affects rabbits [40]. There are plenty of other examples of this phenomenon, though it is not an invariable rule: some viruses are as exacting and specific in their requirements in cultures as they are in nature and must have tissues of their proper host to grow in.

A tiresome but interesting consequence of the widespread use of tissue-culture is the revelation that apparently normal tissues may contain latent viruses which may grow out and destroy the cultures and badly confuse the issue for the virologist. Dr Hull, of Indianapolis, has particularly studied the viruses turning up in monkey kidneys and has described more than forty belonging to several different virus-families [85].

Tissue-cultures can be manipulated and modified in all sorts of ways; they can be combined with the techniques of biochemistry, histo-pathology and electron-microscopy and so we have come to learn some of the details of a virus's life within cells, the next subject to be considered. Viruses have evolved several ways of entering a cell. The most remarkable is that of bacteriophages (phages) attacking a bacterium. Many phages have tails of complex structure and hexagonal heads (plate 12). The heads are made of protein, and their nucleoprotein, which is usually of the DNA type, is contained in a coil within. The tail makes specific contact with chemical substances – receptors – on the bacterial surface (plate 12), fibres within the tail then unwind, an enzyme is released which bores a hole through the bacterial wall and the phage nucleoprotein is injected into the bacterium through the tail. The phage head with all the phage protein is expendable and is left outside: only the nucleoprotein enters to carry on and produce the next generation. There are, however, other phages for which the train of events is not quite the same as this.

A good deal is known of how influenza and other myxoviruses attach themselves to red-blood cells (plate 12) and how they react with receptor substances on the cell-surface in ways manifested in the haemagglutination test. An enzyme on the virus surface breaks down a mucoprotein-receptor on the cell surface and virus is subsequently released again or eluted. It can then attack more receptors, 'browsing' as Burnet calls it, on the cell-surface. Something like this seems to happen when a myxovirus makes contact with a cell of respiratory epithelium it wishes to infect, except that the object of the attachment would of course not be served if the virus were immediately released again to the exterior. Specific receptors probably exist on the surfaces of cells susceptible to attack by other viruses, but less is known of the chemistry of the reaction. Viruses then apparently gain admission to the interior of the cell by a process of engulfment. All sorts of cells, not only the professional phagocytes, can take up small fluid droplets into their interiors. Viruses are probably engulfed in a similar way and are next found in little vesicles within the cell-cytoplasm. There is no evidence that the protein is wholly left outside as with the phages. What probably happens is that the virus causes the cell to produce a stripping enzyme. This very immodestly removes the protein coat and reveals the naked nucleic acid beneath, and this, as with the phages, carries matters on to the next stage. Two other methods by which virus penetrates cells must be mentioned.

Plant viruses may be introduced by the stylets of an aphid directly into the cytoplasm of a cell (plate 31). Again, virus may pass from one cell to another by means of minute intra-cellular channels. This is often seen in members of the herpesvirus group, where foci in tissue-culture enlarge as a result of such a process, although free virus in the fluid part of the culture is hard to demonstrate. Such cell-to-cell transport may occur in the presence of antibody in the outside medium; this cannot get in and stop the trouble any more than a policeman can enter a building and stop suspicious goings-on without a warrant. When it has got inside, the virus enters what is called an 'eclipse phase' in which for a while infectivity cannot be demonstrated by the usual techniques. The virus nucleic acid however, once within the citadel, seems able, by an effective *coup d'état*, to take control of the whole synthetic mechanism within the cell, to halt the cell's normal processes and to divert its energy-providing mechanisms for the production of more virus.

It may enter the cell-nucleus and virus synthesis may begin there, or the whole process may be carried on in the cytoplasm. A most interesting feature with some viruses is the apparent separate production of the nucleic acid and protein components of the viruses [25]. Methods are available, by means of specific antibodies conjugated to fluorescent dyes, for staining separately the virus nucleoprotein and the superficial proteins of its structure. These methods show that in the case of one myxovirus, fowl-plague, the nucleoprotein antigen is made within the nucleus and the other protein in the cytoplasm. With other viruses, too, such as that of herpes, virus is largely synthesized in the nucleus, then completed as it passes along the production-line towards the periphery of the cell. In other instances the different parts of virus-synthesis cannot be so readily separated. In certain instances imperfect virus may be formed, either when the virus grows in unsuitable cells or at abnormal temperatures or in the presence of certain substances which adversely affect one stage in virus-synthesis. For instance empty capsids may be formed, lacking any nucleic acid within. What it comes to is that the cell as a virus-factory is far from completely efficient and needs overhauling. It may make too much of one component of the product and too little of another. Surplus protein may sometimes be found lying around in cytoplasm or nucleus. Also, as we saw in the last chapter, defective manufacture may lead to production of virus-particles of the wrong shape.

The remarkable way in which viruses are replicated points to two possible results of interest when a cell is simultaneously infected with two viruses. There may be a shuffling and fresh combination of ingredients of the nucleoproteins of two viruses multiplying within the cell-nucleus, and one will then obtain true hybrids or recombinants which will breed true since genetic material is concerned [33]. A more frequently observed occurrence follows when a virus nucleoprotein on reaching the cytoplasm acquires the protein coat properly belonging to another virus or a coat of composite or mosaic make-up. It will then have properties depending on the nature of its coat, but its genetic composition will be found to be unchanged when the next generation is produced, so there will be a return to the *status quo*. We do not know how far such goings-on take place in nature and affect the evolution of viruses.

The production of more virus, starting with nucleic acid only, sets viruses apart from larger organisms. With them, it is true, the essential genetical material is still nucleic acid but this is carried on chromosomes which are within the nucleus which is within the cell and all these things divide in a great combined operation when cell-multiplication occurs. The fundamental importance of nucleic acid for viruses was finally demonstrated when it was found possible, first with a plant virus, later with some animal viruses, to infect a cell and produce more virus by infecting a cell with what appeared to be pure nucleic acid. This has been achieved both with riboviruses, including polio and some arboviruses, and also with deoxyviruses such as polyoma. Infectious ribonucleic acid, unlike that which is protected by inclusion in a virus particle, is destroyed by the enzyme ribonuclease, but lacking the appropriate antigenic protein, it is not inactivated by antiviral antibody. Oddly enough it is not so limited as is host-virus by consideration of host-specificity and can sometimes infect cells of normally insusceptible species. This power is, of course, only a temporary acquisition; new virus generated from the nucleic acid is complete virus with normal properties. The virus-protein is seen to be very useful to the virus, for it not only protects it from the ribonuclease but enormously increases its efficiency in entering a cell, even though it does not encourage adventurousness in exploring new kinds of cells. Viruses get out of cells as water does out of a tank, by an orderly trickle or by a cataclysmic burst. The classical example of the second method is bacteriophage. When the phage population has been well built up inside a bacterium, the bacterial wall dissolves, apparently as a result of an enzyme generated in the cell under the stimulus of phage-growth, and the new crop of phage particles is liberated. Adenoviruses, which develop in cell-nuclei, tend to accumulate there and only seem to get out into the cytoplasm when the nuclear membrane is broken and into the outside world when the cell as a whole falls to pieces. Herpesvirus on the other hand, though it too is largely formed in the nucleus, passes quietly out into the cytoplasm and is later released from the cell surface without necessarily destroying it. Of viruses multiplying wholly within the cytoplasm, poxviruses seem to escape mainly when the cell disintegrates but also by passing directly into adjacent cells. With picornaviruses, too, escape of virus seems to coincide very closely in time with cell destruction.

It seems to be particularly the viruses contained within membranes which escape by trickling out a little at a time. They may do this right at the cell-surface or after being first liberated into vesicles in the cytoplasm. Normally there may be little protrusions of cytoplasm at a cell-surface and virus perhaps gets into them, later gaining the exterior as the protrusion is nipped off at the base. In consequence, some viruses are found to have host-components incorporated in their surface-structure. Influenza and some other myxoviruses may stretch out cell-protrusions to an inordinate length before any nipping-off occurs, and thus, with some strains, we find the virus occurring not only in spherical forms but as long filaments. Release of virus by mechanisms such as this may go on for a long while and the cell may survive, being able to take all this in its stride. Often, however, viruses produce destructive effects in cultures or in the living host; the diversion of its efforts towards virus-production then seems finally to have exhausted the cell, so that it disintegrates or shrivels up and dies.

There is one quite different consequence of virus-cell interaction, seen amongst the bacteriophages. This is the phenomenon of lysogeny or production of a lytic agent [77]. 'Lysogenic' bacteria may grow quite normally and seem to be uninfected by phage, yet they may be constantly liberating a phage which is active not on themselves but on other bacteria. Workers in the field call the susceptible organisms 'indicator strains'. The lysogenic bacteria carry the phage in a latent state, in a manner so closely integrated with their own genetic material that it must be attached to their chromosomes, dividing synchronously with them. In this state it is called a prophage. In some cultures individual bacteria will turn up which *are* susceptible and then some mature phage will be liberated, detectable by the use of an indicator strain of organism. At times not even this happens, but production of mature phage from prophage can be 'induced' by certain manipulations; ultra-violet irradiation is one of the most successful. When this is done it could readily appear to anyone ignorant of the background that the irradiation had brought an entirely new virus into being.

Some phenomena amongst animal viruses have a resemblance to those of lysogeny, prophage and induction in the bacterial viruses, but no one has yet been able to show that there is an exact parallel.

One particular result of the activity of viruses within cells requires mention – the formation of so-called 'inclusion bodies'. These may be 'intranuclear' or 'cytoplasmic' (plate 16). They may consist of masses of virus particles, minute colonies of virus; or they may be composed of something else or develop as a result of damage done by the virus. Minute virus-colonies form inclusions in the cytoplasm in the case of many poxviruses, but there are other inclusions in these infections composed of other materials. Some adenoviruses and picornaviruses occur in regularly arranged 'crystalline' masses, those of adenoviruses in the nucleus, those of picornaviruses in the cytoplasm. There are also, in the case of adenoviruses, masses of smaller particles, possibly surplus protein not required for making virus. Of special interest are the intranuclear masses occurring in infections by members of the herpesvirus group. These usually stain pink with eosin and other acid dyes and are separated by a clear halo from other nuclear material which collects at the nuclear surface and stains blue. This appearance is an artefact due to the way in which material has been fixed for staining, but it presents nevertheless a very striking appearance and is often useful in diagnosis. Though the herpesviruses apparently start their development in the nucleus, they have mostly left to enter the cytoplasm by the time these inclusions can be seen; they have been described as gravestones; if so they seem to be inscribed, so characteristic are they, with the name of the murderer of the cell.

Little need be said at this point about the various ways in which viruses behave in relation to the whole host; this will be better dealt with when we come to consider individual viruses. Some viruses, it will become clear, enter the body by a particular route and cause all their damage at or near the portal of entry. Wart-viruses get in through the skin and that is where they produce their effects, quite locally. Common cold and other viruses enter by the respiratory passages and attack the mucous membranes there, not quite so locally as does the wart virus but affecting only the tissues close to where they made their original landing. The story is more complicated in the case of many viruses which produce a generalized infection. Many of them have a special affinity for some particular part of the body, often referred to as a 'target organ'. Thus, mumps enters through the respiratory tract but does not do any obvious immediate damage there. Where

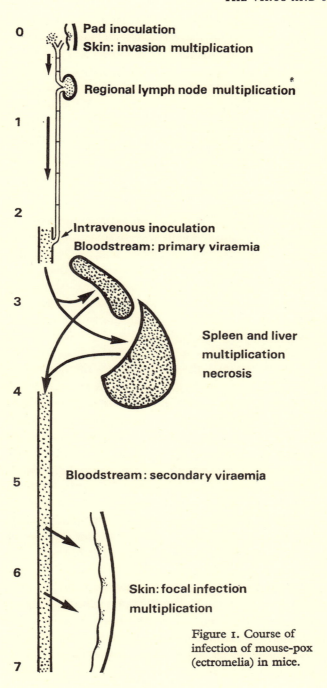

0 — Pad inoculation
Skin: invasion multiplication

Regional lymph node multiplication

1

2

Intravenous inoculation
Bloodstream: primary viraemia

3

Spleen and liver
multiplication
necrosis

4

5 — Bloodstream: secondary viraemia

6

Skin: focal infection
multiplication

Figure 1. Course of
infection of mouse-pox
(ectromelia) in mice.

7

it first multiplies we are not sure, but after an incubation period it gets into the blood-stream and from there reaches its favourite target organs, the parotid and other salivary glands. It has other, less frequently favoured, targets, the testes, ovaries, pancreas, the meningeal membranes round the brain and some other sites. Many viruses, those which cause rashes or exanthems, have the skin as their target organs. With members of the pox group it seems that a little local multiplication of virus may occur at the point of entry, perhaps the respiratory passages or the skin, but there is soon spread to the local lymph-nodes, further multiplication there and thence a spill-over into the blood-stream and a settling in the skin to produce the characteristic pocks [57]. Other organs may be affected also, for instance the lungs when smallpox is complicated by pneumonia.

Of particular interest are the viruses which affect the central nervous system. Such attack is usually an incident in an infection, often quite uncommon and not at all helpful from the virus' point of view. An exception is rabies, in which the infection of the brain leads to the excitement which causes a mad dog to bite its victim and spread the disease: this is made possible by presence of the virus in the other target organ, the salivary glands. The encephalitis and other forms of damage to the nervous system are sometimes the result of attack on blood-vessels, destruction of the nerve-cells being secondary to this: in such cases the virus has of course been spread by the blood-stream. In other instances the nerve-cells of cord or brain may be themselves the targets, as is seen in poliomyelitis. There is evidence in the case of some such infections that virus reaches the central nervous system by passing along nerves centripetally from a peripheral site. This seems to be true for infections of rabbits with herpes virus, and in the case of rabies. It was formerly thought to be true for poliomyelitis and may be so to some extent, but transport of virus through the blood is now considered more important. There has been much dispute, where passage along nerves does occur, as to whether it is actually along the nerve-fibres themselves or in spaces between the fibres. The view mainly held at present is that transport in tissue spaces certainly accounts for some instances of passage along nerves, but that passage along nerve fibres is not yet proved conclusively.

When arboviruses multiply in insect vectors, they do no known harm to the insects; but transmission to a vertebrate victim must take place if the race of viruses is to be carried on; so virus has to reach the salivary glands to be injected into the vertebrate victim when the insect bites. So mosquitoes' salivary glands can be considered target organs, even if the infection in them is an inapparent one.

Plant viruses may produce infections which remain local or which become systemic. Spread occurs from an infected cell to adjacent ones through communicating protoplasmic channels. Spread when it is systemic is by way of the plant's phloem, the phloem cells themselves being undamaged. It has been suggested that where infection remains local, there is damage to phloem cells so that spread by that route cannot occur. Some plant-viruses cause systemic disease in some species of plants but only local lesions in others.

Chapter 4

Defence Against Viruses

WE SHALL LEARN in the concluding chapter that control of viruses by anti-biotics and other drugs is a study in an early stage of development, in contrast to the great advances in the use of such substances in dealing with bacterial infections. There is, however, something to be said on the credit side: immunity after attack by many viruses is life-long. We all know that chickenpox, measles and mumps are diseases of childhood, once suffered, over and done with. Second attacks are rare. An attack by most viruses leads to the production by the body of modified globulins in the blood-plasma: these are specific neutralizing antibodies which play an important part in immunity. They do not appear to kill the viruses directly, for, at least for a time, an antibody-virus combination can be split up again with restoration of the virus's activity. They do, however, affect the physico-chemical properties of a virus and block the 'business-end' of it which permits it to enter a cell. Antibodies seem to be unable to enter a cell, so, once a virus has got inside, it can go ahead and snap its fingers at antibody outside the cell. Antibodies are thus often quite effective in prophylaxis, in preventing infection, but of little value for treatment once infection has started. Neutralizing antibodies may persist for a very long time. Sera of old people in the southern United States were found still to have yellow-fever antibodies, seventy-five years after the elimination of that infection from those parts; there could have been no contact with it in the meantime. Immunity may also be kept up by the repeated stimulus of contact with small doses of virus. Old people, away for a long time from contact with infections such as measles and mumps, sometimes catch the infections from their grandchildren. Usually, however, immunity is very good against the diseases with long incubation periods. We saw in chapter 3 how an infection such as that due to a poxvirus builds up in several stages – local lesion, then lymph-node involvement, then spread through blood-stream. Even if immunity has waned a little and is incomplete, there is time enough to mobilize the defences rapidly when the infection is only at an early, inapparent stage. The virus is halted before it reaches the spreading stage and the subject never knows he has been infected at all.

All this is very satisfactory, but unfortunately, as we all know, it does not apply to all viruses. We can have colds, we can have flu, over and over again. The immunity may fail to work in such cases for one or other of three reasons. Where there is a short incubation period of only two or three days, the infection does not develop in stages, so an immunity which has waned somewhat does not have time to develop in time to stop all symptoms. Again, when an infection involves only superficial membranes, such as those of nose and throat and does not involve spread by the blood-stream, defence through antibodies cannot work so well. True, some antibody gets into the mucus covering these membranes, but it is in much lower concentration than in the blood and is therefore at a disadvantage. Finally, many viruses exist in a number of different antigenic types; this means that their surface structures are sufficiently diverse in their chemical make-up that antibody against one does not neutralize the others or only partly and imperfectly.

There are numerous antigenic types of rhinoviruses, echoviruses, adenoviruses and the foot-and-mouth diseases viruses. It may well be that against these viruses specific immunity is rather better than was once thought, its apparent failure being due to the existence of all these different virus-types. As we shall see in chapter 6, the virus may not achieve its success because there are so many already existing types; it may, as does influenza, manage its repeated come-backs by a remarkable power of variation, making its own new antigenic types between one major outbreak and another.

In yet a third class of virus-infections, the virus somehow fails to generate effective antibody, so that virus is never eliminated but is liable to cause relapses or recurrent attacks or to persist as a chronic blood-infection. Herpes simplex does generate good neutralizing antibodies, but, nevertheless, persists within some cells, probably spreading from cell to cell unhindered by antibody, and breaking out from time to time from the restraints which control it (see p. 163). The viruses of human hepatitis and of equine anaemia may be present in the blood-stream for months or years. Viruses within tumours such as the Rous chicken-sarcoma are of course very well placed, for their own multiplication within the safe harbourage of cells is favoured by the multiplication of the cells which they themselves are stimulating.

There has been considerable argument as to whether, in instances where immunity is enduring, it may not be maintained by the persistence of virus somewhere in the body, tucked away where it cannot get out and produce overt disease or infect other people. It is known that after some infections such as those of vaccinia and foot-and-mouth disease, virus may be recoverable for many months. It has been suggested that, in contrast to the

35

examples just given, where continuing or relapsing disease may occur, virus may in some cases persist in a modified, imperfect, non-infectious but still antigenic state. Until lately such theorizing had no solid basis: recently, however, studies of tumour-producing viruses have shown that just that very state of affairs may occur (see p. 192).

We have discussed up to now the mechanisms of defence based on orthodox immunology. There is, however, an entirely different way in which virus-infection may be halted, through the activity of interferon [87]. It has been known for a long time that infection of cells with one sort of virus may protect them from infection by another, perhaps quite unrelated. The phenomenon has been called interference and in many instances it is mediated by interferon. This is a protein of rather small molecular weight, made within cells and therefore able to stop a virus in its intracellular fastnesses where antibody cannot penetrate. It is produced and can act much more quickly than antibody and serves in fact as a first line of defence while antibodies are being mobilized. Its production by the cell is called forth in response to the stimulus of dead or damaged virus, or perhaps by any foreign nucleic acid, and it is effective to varying degrees against viruses of different families, not merely against the one which called it forth. On the other hand it is present only transiently, though that may not matter if it holds the fort till the antibodies can take over the defence. Also, though not specifically directed against any particular virus, it is specific in the sense that interferon produced by the cells of one animal species is only effective, with minor exceptions, in the same species: this fact limits the possibility of producing interferon on a large scale for treatment of virus infections. There are suggestions in published work that as a rule virulent strains of a virus elicit less interferon or are less readily inhibited by interferon than are avirulent strains. It also appears that interferon is produced less readily and operates less readily in many cancer cells than in normal ones.

In discussing antibodies, we have so far only considered those which neutralize the activity of viruses, and these are indeed the ones of chief importance in the body's defence. Tests for their presence are made by mixing virus and antiserum, keeping these in contact for a suitable time and inoculating the mixture into an experimental animal or into a tissue-culture and seeing whether the virus's activity has been prevented. This is often a less-than-perfect method; others may be simpler, cheaper and less time-consuming. Other 'serological' tests for presence of antibodies in sera are usually available. Though the techniques used are different, the virologist is often, but not always, measuring different manifestations of activity of the same antibody.

A test of great usefulness is the haemagglutinin-inhibition test; this is of course only applicable to those viruses which clump red-blood cells or haemagglutinate. It was first discovered [78] for influenza, but it has since been found that many viruses agglutinate cells of some vertebrate or another; it may be necessary to hunt for a species with susceptible red cells and to fiddle around with pH and temperature until one finds something that works. For many arboviruses, goose cells are more sensitive than any others: for vaccinia and some other viruses, fowl cells and only cells of particular fowls, are most suitable.

The substance which clumps the red cells, the haemagglutinin, is usually on the surface of the virus particle; it may be separable from it as a free, much smaller, particle; or it may be so closely bound to the virus that only the whole virus acts as a haemagglutinin.

To perform the haemagglutinin-inhibition test, measured small volumes of the virus and serum under test are placed in small cups in a plastic plate; red cells are added and after a suitable interval the way in which the cells have settled is recorded. In the absence of virus or when its activity has been neutralized by antiserum, the cells settle into a compact clump at the bottom of the cup; where the virus has been able to agglutinate the cells, these settle in a diffuse pattern instead. By making serial dilutions of either virus or serum, one can measure each of them quantitatively, with considerable accuracy. This test is all important in the separation of closely related viruses, for instance in deciding whether a newly isolated influenza strain is really different from older ones. The haemadsorption test [177], referred to on p. 17 is really a modification of the haemagglutinin test, suitable for viruses which do not clump cells quite so readily; and an inhibition test with antisera is applicable also to measure antibodies against haemadsorbing viruses.

When antibodies unite with viruses they may clump these, though this test requires such a lot of virus that it is not of great practical value. When they unite with soluble antigens they may form precipitates with these. The antigen-antibody complex can be shown to bind to itself a complex but useful substance present in fresh sera, especially guinea pig sera, and this is the basis of a 'complement-fixation test' which is useful for many purposes. Again, antigen and antibody can be placed in little cups cut out of a sheet of agar-jelly in a Petri dish. The substances diffuse through the agar and where antigen meets antibody a white line of precipitate forms. Some viruses such as poxviruses contain a number of antigens in their composition. These may diffuse at different rates and so produce a number of separate lines. Tests can be put up with antiserum to one virus in the middle cup and different viruses in the outer cups. Three or four, or even as many as seventeen lines [182] may be seen where virus reacts with its own antibody, but smaller numbers when related viruses are concerned. Such a test may show that two viruses have some antigens in common but not all; the viruses are therefore related but not identical. The test can be referred to as the gel-diffusion test.

Another very useful test is the fluorescent-antibody test, first described by Dr A. Coons of Boston [44]. Antibody can be attached by chemical manipulations to a fluorescent dye. Such a 'tagged' antibody can be applied to a thin section of a virus-infected tissue and examined under the microscope. The tagged antibody will unite with any antigen present and microscopical examination will then show fluorescence wherever the antigen plus antibody lies. The test is particularly useful for showing where antigenic viruses lie hidden, in what tissues of the body and whether in the nucleus or cytoplasm of the cell. Sometimes such a technique is the only one available for detecting presence of a virus – if it happens to be one producing no changes in inoculated animals or tissue-cultures and not detectable by the other serological tests.

Part 2
Transmission of Virus Infections

Chapter 5

Catching Colds

WE CAN BEST begin a discussion of how virus infections are transmitted by considering the minor infections of the respiratory passages, that is the nose, pharynx, larynx and trachea. The commonest and hardest to deal with of these is the common cold.* It is important to remember that a number of viruses belonging to different families can cause colds, sore throats and troubles of that sort. The rhinoviruses seem to be the commonest of the known culprits, but the para-influenza viruses, adenoviruses and others play a part too. The symptoms produced by these agents are far from being characteristic. A running nose or a sore throat may be due to one or other of several viruses. One cannot say which on purely clinical grounds, at least with any certainty, though there is a *tendency* for rhinoviruses to cause running noses without fever and for adenoviruses to give rise to pharyngitis, usually with fever and sometimes with conjunctivitis too; and so on. Influenza gives rise, as we all know, to fever with aches and pains, but that we shall leave till chapter 6.

The pattern of attack on populations differs also, as we shall see. Most of what needs discussing in this chapter concerns these infections in general, but particularly colds caused by rhinoviruses.

Rhinoviruses are picornaviruses, viruses of the smallest, simplest kind, adapted to grow in the nose rather than in the intestine, which is the normal habitat of other picornaviruses such as polio and other enteroviruses. This adaptation is reflected in the rather different requirements of the rhinoviruses for growth in tissue-culture: they prefer a more acid medium than do enteroviruses, a rather lower temperature and plenty of oxygen. These are the conditions they find in the nose. Further, though they like a slightly acid medium to grow in, they are killed by higher degrees of acidity such as do not bother the enteroviruses at all. This again makes sense: the enteroviruses have to reach their target organ in the intestine and to get there they have to pass through the stomach with its acid gastric juice; the rhinoviruses

---

* The whole subject of colds is dealt with at greater length in another book, intended for laymen and others: *The Common Cold* by C. H. Andrewes, Weidenfeld and Nicolson, London, 1965.

have no such need and in fact are rarely if ever found in the faeces, doubtless having been killed off by gastric acidity.

A good deal of what we know about rhinoviruses has come to light in the course of work at the Medical Research Council's Common Cold Research unit at Salisbury, with which I have been associated since its beginnings in 1946. Facts about colds have had to be laboriously uncovered by means of experiments on human volunteers, of whom more than eight thousand have visited the Salisbury unit during twenty years. Only since the discovery in 1959 of how to grow rhinoviruses in culture has it been possible to make really satisfactory progress [174].

It might well be supposed that the natural history of colds was a simple matter. The infection is one of those in which production of symptoms, coughing and sneezing, helps the spread of the virus: if it were wholly symptomless it might not be able to spread satisfactorily. A virus-carrying sneeze infects another subject and so the infection is carried on. All very simple and straightforward but unfortunately not fitting all the facts.

Colds of course can be transmitted experimentally by dropping infected nasal washings up the nostrils of another subject. It is also true that a cold may be caught by a person in close contact with a 'cold' sufferer; it has been shown experimentally that this happens. It seems, however, that the infectivity is rather low. Children have been shown to spread infection more readily than do adults; perhaps it is because their personal hygiene is rather less perfect or perhaps virus grows more abundantly in their noses. Yet in experiments carried out at Salisbury only ten per cent of adults exposed at close range for several hours contracted colds from children with colds [100]. This was discovered in an experiment in which children's parties were held; the qualification for being invited was that the child must have a really good cold! Epidemiological studies in the neighbouring Chalke Valley showed that the chances that an adult would contract infection from a cold in the same household were only one in five. In similar investigations in offices in London and Newcastle Drs Lidwell and Williams [99] found that one was rather more likely to pick up a cold from someone with whom one worked at close quarters than from those more remote. Nevertheless in a majority of cases people with colds could give no history of a contact from whom they were likely to have caught the infection.

Epidemiological evidence of another sort makes the idea of simple catching an unlikely explanation. Workers in Holland, the USA and elsewhere have recorded that outbreaks of colds are likely to appear simultaneously over wide areas in a way impossible to reconcile with ordinary person-to-person spread. Van Loghem [175] plotted curves of incidence of colds obtained from various parts of Holland. Not only did they coincide as regards the time of occurrence; they were also remarkably similar as regards the height of the curves, a fact showing how great was the similarity in the proportion of people affected in the different areas. The waves of colds were related to changes in weather. Before, however, we deal with that side of things let us look at immunity to colds.

We all know that it is possible to have colds again and again, often several in one winter. Rhinoviruses cause only superficial infections of mucous membranes and that may be one reason, as discussed in chapter 4, why immunity is not permanent as it seems to be in mumps and measles. There is, however, another, probably more important cause of trouble. There are now known to be many rhinoviruses. These have been divided into M-strains which grow in monkey as well as in human tissue cultures and H-strains multiplying only in cultures of human origin. When neutralization tests are put up, to discover whether a particular serum will knock out a particular virus in these cultures, it becomes clear that there are very many antigenic types of these viruses. We do not yet know how many: there seem to be at least eighty. Many of these have turned up on both sides of the Atlantic and antibodies to some of the M-strains, which are easier to study, are present in sera from all over the world. But the search is only just beginning and we cannot guess how many different types may not turn up within the next few years. A good level of antibodies to one type seems to be related to resistance to that type and such resistance can be conferred, we think, by vaccination; unfortunately there is good reason to believe that immunity to one of these viruses gives little or no protection against other types. One worker in the field had three colds over a period of several years; one of them yielded an M virus, one an H virus, while no virus could be cultivated from the third cold [173]. Dr Dorothy Hamre of Chicago tells me that she isolated five different antigenic types of rhinoviruses from five colds occurring in one individual in the course of eighteen months; moreover in her studies generally she never isolated the same virus twice from colds in the same person. Two other workers in Chicago, Drs Jackson and Dowling [88], studied five strains of viruses and obtained evidence that immunity after inoculation with a cold virus was good for at least many months to that virus, though not against the other four. It is possible, therefore, that immunity to cold viruses is not as transient as we used to think; it may be the multiplicity of viruses which is the main cause of trouble. We shall learn more in chapter 6 about how influenza viruses keep changing their antigenic make-up and thus coping with an immunity in the

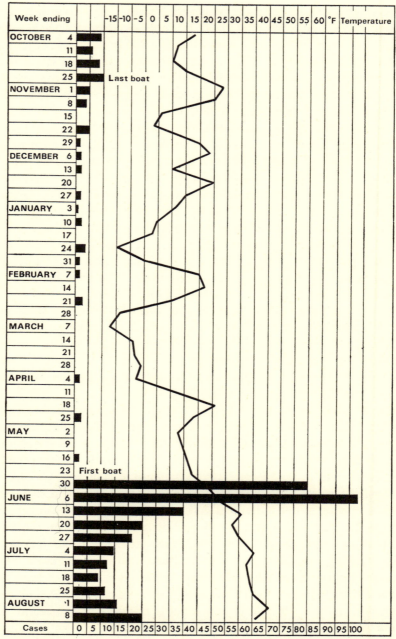

Figure 2. Incidence of colds in an isolated community (Spitzbergen).
Vertical columns indicate number of colds each week. Solid line
indicates temperature (°F).

44

population. We have as yet no idea as to whether rhinoviruses change in a similar way or whether there are just very large numbers of stable types.

Besides specific immunity there seems to be resistance of another sort directed against colds generally. Lidwell and Williams [99] found evidence of some degree of resistance for a few weeks after a cold; one would not expect any at all if resistance to all the cold viruses was varying independently. It is very well attested that people in isolated communities, particularly polar explorers, are free from colds while in isolation [124], but are likely to go down with a severe one when they once more make contact with the outside world (see figure 2). There is no reason why this should be so, if there were a reasonably long-lasting specific immunity against all the viruses concerned. The facts can, however, be explained if there is some degree of at least low-grade generalized anti-cold immunity. This might well be maintained in larger centres of population by constant stimulation with small doses of various cold viruses. It is a familiar experience that one is liable to catch a cold soon after return from a health-giving outdoor holiday. This may be due in part to the fact that returns from holidays often coincide with the onset of colder autumn weather. It is likely also that immunity may wane after even a few weeks freedom from the constant interchange of viruses which is probably happening in town. Nowadays one tends to speculate as to whether interferon may not be concerned in this mechanism of non-specific resistance.

When people talk of catching a cold they mean one or other of two things: that the infection has been passed to them from another person or that the cold has been brought on by chilling. There is a very widespread belief that the latter is a common happening. It may be true, though there is no scientific evidence for it. In experiments at Salisbury we tried to produce colds in volunteers by chilling them in various ways: we failed. When we combined the chilling with the administration of a very small dose of virus, we obtained no evidence that the chilled volunteers were any more susceptible than controls who were not chilled [4]. Dowling and Jackson in Chicago [50] obtained similar results. There was one difference, however; women in the middle third of their menstrual cycle did seem to be more likely to catch colds when chilled. Our failure to show an effect of chilling on susceptibility does not prove that such an effect is non-existent: it may only mean that we did not get the experimental conditions right. As we shall see later, there is a possibility that chilling is one form of stress which may predispose to colds: our volunteers may have been insufficiently stressed; indeed they regarded the whole thing as rather fun.

Discussion of chilling leads naturally on to the question of why colds are commoner in cold weather, and this is a really complicated and puzzling problem. Obviously cold weather alone does not cause colds, or those isolated polar explorers would not be so free from the pest. It is a general experience that colds are less frequent in summer, that there is often a wave coinciding with the first cold weather of the autumn, often another wave about the New Year perhaps grumbling along with some increase in activity in March. It seems possible that the various cold-causing viruses may prefer different seasons for their greatest onslaughts: rhinoviruses are very prevalent in the October waves while myxoviruses, such as the para-influenzas, are more to the fore in January. Influenza itself rather favours January: epidemics of flu are more apt to begin about the New Year than at any other time, though there may be first signs of something in December.

Figure 3. Relation between temperature and incidence of colds in a medical practice. The colds (heavy line) vary with temperature in the earth (thin line) to show up the relationship. The temperature has been inverted so that colder temperatures are at the top of the chart.

A number of attempts have been made to relate incidence of colds to particular states of the weather, such as cold, humidity and rainfall. Dr Hope-Simpson, a general practitioner in Gloucestershire, got some of his patients to keep records of when they got colds [79]. He was able to plot the resulting figures as a curve (figure 3). He found an extraordinarily good correlation with fall in temperature; he took the temperature a foot down in the earth, thus evening out transient fluctuations. In the figure the curve of temperature is inverted, so as to show better how closely the curves of 'cold' and 'colds' fit each other. Hope-Simpson thought that the effect of temperature was not a direct one, but that the colder weather caused people to light their fires indoors and begin to live in conditions of lower relative humidity. Humidity indoors soon begins to fall lower than outside, as soon as fires are lit. This could be an important factor: on the other hand so

46

many things change as summer turns into winter that it is very hard to say which factor is the important one. It could be the cold itself or its effects on one's metabolism or the results of lighting fires or changes in social habits or putting on more clothes or perhaps something no one has even thought of. One theory, not very likely to be true, has been that taking out winter overcoats shakes cold germs out from their summer-time hiding places.

Various theories as to how winter favours colds can be divided into two main categories: there may be an effect making cold viruses more easily transmitted from one person to another; or there may be a lowering of people's resistance. A view was put forward by some Dutch workers [76] in support of a possible effect in the first category. Influenza is mainly a winter disease, while poliomyelitis is commoner in summer and autumn. Particles of influenza virus sprayed into the air in the form of a very fine mist were found, over a certain temperature range, to survive better when the relative humidity was low, as it is in winter indoors. On the other hand the polio virus retained its infectivity better when the relative humidity was rather higher. Thus one virus might have better chances of passing from one person to another in the winter and the other in summer. A good many objections were raised against this theory. One in particular seems to be rather damning. Rhinoviruses, though they may be isolated in summer, seem, like influenza, to cause more colds in the winter, yet in most of their properties they resemble their fellow picornavirus, polio. It was in fact shown by Buckland and Tyrrell [27] at the Salisbury unit that rhinoviruses survived better, as did polio, when the air was *not* too dry. This does not at all fit in with the idea that a dry atmosphere encourages the spread of colds in winter.

Most workers agree that no particular state of the weather, temperature, humidity, wind or rainfall predisposes to colds. The thing that matters seems to be a sudden change. This could operate better on people than on virus; but it still might do so either directly by affecting their resistance or less directly by making them change their habits and making them crowd together and therefore interchange their viruses more freely. I have never felt much attracted to the idea that crowding was all that important; for people in towns get packed closely together all round the year especially in public transport. Quite a small seasonal change in weather is said to bring on outbreaks of colds in a West Indian island [116], changes which in a more northern climate would be without effect and regarded as trivial. In tropical parts of Asia waves of colds are more apt to come at the start of the monsoon rather than with any fall in temperature. Colds are also said to start up amongst passengers when a liner passes from the tropical Indian Ocean to the cooler Mediterranean.

Before trying to put together the pieces of this jig-saw puzzle, let us consider by what mechanisms rhinoviruses and other agents could get from one person to another. Even though such transmission may not explain everything, we have to admit that these viruses do, somehow, get passed about. There are three well-recognized vehicles for transmission of respiratory infections (see figure 4). Particles are ejected with considerable force from the nose and mouth during sneezing and coughing and even loud talking, and these may travel for considerable distances, about fifteen feet, before falling

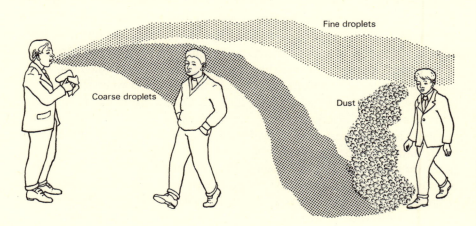

Figure 4. Possible ways of transmitting colds; by coarse droplets reaching a man who is close to; by fine particles remaining suspended long enough to reach a man farther away; and by coarse droplets settling to be dispersed later as dust.

to the ground. They may thus pass readily to another person at close quarters: we call these 'coarse droplets'. Rather smaller droplets shot out at the same time evaporate almost instantaneously so that nothing remains but the minute solid particles which the droplet contained. These are called 'droplet nuclei' and may well contain germs. They may float in the air for up to an hour and might thus infect people at greater distances than could coarse droplets. Finally, the material from coarse or fine drops, having reached the ground, may dry up and be redispersed later as dust, when, for instance, rooms are swept or beds are made. An attempt was made, at Salisbury, to settle which of these methods was most important in the case of colds. As transmitters of infection we used the children with colds coming to the special children's party mentioned on p. 42. Most colds were transmitted when contact was close enough to permit coarse droplets to reach their targets: but only ten per cent of exposed people caught colds even then after four hours' contact. When a blanket was hung from the ceiling, forming a barrier to such direct transmission but allowing droplet-nuclei to drift round its edges, some cross-infections still occurred.

A number of experiments suggested that transmission in dust or by manual contamination was unlikely to be important. Handkerchiefs used by people with colds were dried in the air and given to normal people to use: no colds resulted. Painting infected secretions on the outsides of the nostrils did not transmit infections. Gauze strips soaked in cold secretions were applied inside normal nostrils; these did pass on infection, but not if the strips were dried first. It seems then that the agents we were studying were easily killed by drying; they were probably rhinoviruses but this was in the days before rhinoviruses had been isolated.

Further experiments were carried out at a later date [28] using another cold virus, Coxsackie A21 (see p. 75); this virus was convenient as it causes a high proportion of colds in volunteers and is easily identified and measured in the laboratory. Very small doses of virus were applied with swabs either just inside the nose, into the eye or in the throat. Virus given directly into the nose was the most effective; that put into the eye did not do quite so well; that in the throat was still worse. Virus in the eye could readily have run down into the nose by way of the duct which carries down surplus tears, secreted by the lacrimal gland. We thus have evidence as to how that virus best got in: how did it get out? Pictures taken by high-speed photography show that when one sneezes, most of the fluid expelled is in fact saliva from the mouth. Volunteers were caused to sneeze into plastic bags and the observers gathered separately the secretions from the mouth and the nose. One of the subjects was very expert at producing a sneeze 'on demand'. Six times as much secretion was collected from the mouth-bag as from the nose-bag. However, the concentration of the Coxsackie virus in the nasal secretions was a hundred times as high as in the saliva from the mouth. So more than ninety per cent of the virus coming out in a sneeze would be of nasal origin. The Coxsackie A21 was a useful indication of what is almost certainly happening in the case of rhinoviruses and some other respiratory infections.

There are some interesting differences in the behaviour of rhinoviruses and some of the others. The para-influenzas 1, 2 and 3 are important causes of minor respiratory infections amongst small children, particularly those living together in nursery schools. They cause not merely colds but feverish illnesses and particularly croup caused by inflammation of the larynx and trachea. Second infections are very frequent; third infections are less so, so

49

perhaps immunity does build up in time. However, reinfections may occur even in adults and then the virus does seem to cause common colds rather than infections of the larynx and windpipe. Even in adults immunity is not very enduring. As mentioned earlier, outbreaks due to para-influenza infections are commoner about January and February. The para-influenza viruses are not a really common cause of common colds in adults.

Of considerable importance is a myxovirus which is very similar to the para-influenzas by electron-microscopy. This is the respiratory syncytial or RS virus, so-called because it forms syncytia in cultures; these are big cells with many nuclei caused by running together and fusion of smaller cells. The RS virus was first recovered from colds in chimpanzees and was then called the chimpanzee coryza agent. It is the commonest cause of the more serious respiratory infections in small children, bronchitis and bronchopneumonia. Second attacks may occur in childhood, but when reinfections occur in adults, they take the form of common colds, just as para-influenza infections do. The RS virus comes along in epidemics just as influenza does and almost every year; the outbreaks come in winter but not always at the same time in each winter. Whenever many cases of bronchitis and pneumonia are turning up amongst small children, it is pretty sure to turn out that the RS virus is being frequently isolated.

The adenoviruses are particularly interesting. As already mentioned there are about thirty antigenic types. Some of them cause minor infections in children and thereafter remain for some time in a latent state in their tonsils and adenoids. Other types, especially numbers 3, 4 and 7 may cause some trouble amongst civilians, though very little. But when service recruits are brought together for the first time, there is very commonly indeed an outbreak of feverish sore throat due to a 3, 4 or 7 type of adenovirus. The Coxsackie A21 virus behaves in a similar way, being very apt to cause colds in the same type of population. Influenza B may do this too; and so may influenza A at times when it is not, as it were, in full spate and sweeping through whole communities. It is quite a puzzle as to why some viruses only get going in these closed communities and not in the population at large. A possible explanation is based on the fact that different people suffering from infection with the same virus may shed many different amounts into their environment. The amount of virus in washings from people with influenza may vary over a million-fold range. Some babies scatter veritable clouds of bacteria into their surroundings: such 'cloud-babies' are a dangerous source of infection to others [53]. If an individual is thus dispersing remarkably large numbers of adenoviruses and so infecting a number of other people, nothing much will be apparent if he is 'at

50

large'. The victims whom he infects on an underground train will be scattered far and wide and their infections will never be traced to a common source. If, however, he is a member of a small community, those he infects will remain together and there will be an evident outbreak of disease.

Another, unexpected factor, has been brought to light recently by observations made at the Great Lakes naval training station in Illinois [127]. Newly recruited men come here for ten weeks' preliminary training before being dispersed to other units. During the first half of this period they undergo immunization against many of the infections they are likely to meet – adenoviruses, influenza, polio, smallpox, typhoid, diphtheria and tetanus. Soon after they arrive at the unit there is very commonly a considerable outbreak of upper respiratory infections to which the adenoviruses largely contribute. The outbreaks affect only these recruits and not the seasoned men at the unit. The experiment was tried of delaying the bulk of the inoculations, to see if that affected the incidence of respiratory infections. All the men received adenovirus, influenza and one polio 'shot' at the beginning of things, as those infections were felt to constitute an immediate hazard. The other injections were delayed till the second half of the ten-weeks' training. The result was consistently the same: delay in giving the vaccines led to a twenty per cent reduction in the incidence of those respiratory infections which were sufficiently severe for the men to report sick. There was a similar difference in the incidence of rubella (German measles). The results were interpreted as meaning that the stresses of the first few weeks' training increased for the time the susceptibility to respiratory infection. The stresses included the reaction to a dramatic change in mode of life with strict discipline, and relaxations in this field helped to reduce the colds and similar troubles. The stress of numerous inoculations seems definitely to have helped to increase the rate of infections. There is no doubt that over a longer term these inoculations greatly benefited the men by reducing the danger from polio and all the other infections; but this benefit was partly cancelled out by making the men temporarily more susceptible to the infections against which they had no specific protection. How the stresses in question operate is unknown; it may be there is something in common between them and stresses imposed by a sudden change in weather to which our adjustments are not quite fast enough.

We can now consider how it may be possible to fit all these confusing facts into any sort of overall theory. It seems certain that infections with cold-viruses can be directly spread from one person to another; also that the efficiency of spread is apparently not very great. It is also obvious that the facts cannot all be easily explained by a simple person-to-person spread such as apparently suffices in the case of measles, where inapparent infections are rare. The explanation of the seasonal character of colds is not so obvious. Could cold and winter operate by activating from time to time a chronic latent infection, just as herpes simplex is activated by infections and other stimuli? (see p. 163). Telling strongly against that is the story that different cold viruses have been isolated from successive colds in one individual. I have tried [9] to reconcile the apparently contradictory facts by supposing that all sorts of viruses are constantly being interchanged by people at a low level of activity: that is to say these assorted viruses may be affecting a temporary lodgement on a person's mucous membranes, but, coming only as spies, not in battalions, they are held in check by local defence mechanisms, perhaps interferon. They thus fail to develop to the point where they either cause disease or give rise to solid immunity. If, however, a suitable stress strikes the person concerned, his local defences may be upset and he may 'catch a cold'. And this cold will be 'due to' whichever of a number of viruses he happens to be harbouring at the moment. There are undoubtedly some difficulties in the way of this hypothesis. Spreading of a detectable respiratory infection needs obvious coughing or sneezing to be effective: one has to imagine that this subtle infiltration by spies takes place without such obvious and dramatic means of spread, and its occurrence may be very difficult to demonstrate.

I suggested earlier in this chapter that constant stimulation by small doses of viruses might stimulate a non-specific anti-cold immunity; and now I am arguing that the receipt of such small virus doses may be at the basis of our liability to colds. These two ideas are not inconsistent: the little doses and temporary lodgement of viruses may stimulate an immunity which is useful but not absolute: it could still be overcome by the appropriate stress. All these infections are of great importance to the community from the point of view of working efficiency and there is scope for much work in the field along many different lines and by people trained in different disciplines. Only thus will this most puzzling of problems be solved.

The behaviour of rhinoviruses illustrates an important principle in the natural history of viruses: a virus must not spread too efficiently. If it does so, it will soon exhaust the supply of susceptible victims – or else must find a way to combat this danger. Things would go very nicely for rhinoviruses if they really behaved in the way I have suggested, spreading as a limited infection not causing symptoms or inducing immunity, but with the potentiality of being activated and giving rise to a sneezing disease. The stress – by weather-changes or other means – would operate only at intervals and thus the viruses would get well spread about over a considerable period of time yet only intermittently. Evolution towards such a state of alternating quiescence and activity might well ensure that the common cold would be never away from us for very long.

Chapter 6

Influenza, Ringer of Changes

WHEREAS COLDS come along every winter, influenza behaves in a more variable and unpredictable way. In some winters it doesn't appear at all, in others it causes widespread outbreaks or perhaps only localized ones; and yet again it has proved capable of causing the most lethal epidemic in history. Accounts of its earlier doings have to be accepted with some reserve, for the virus was not isolated and made available for laboratory study until 1933 [157]; so we are not quite certain that what was called influenza before 1933 was all due to the same virus; nevertheless there is a very strong presumption that this was the case [163].

During the first half of the nineteenth century there were a few fairly big and widespread outbreaks of flu, but after 1850 the disease apparently disappeared from most of the world for a period of nearly forty years. In 1886 and 1887 some influenza was recorded in Russia; apparently its virulence was exalted in Bokhara in Central Asia in the summer of 1889 and later in that year it spread to other parts of Russia and to western Europe. So began the world-wide epidemic, or rather pandemic, of 1889–90. Second and third waves caused progressively higher mortality. The remarkable thing about the outbreak was that it seems to have started something and a good deal of influenza has been with us ever since. As the epidemiologist Greenwood wrote 'the position lost has never been regained'. Figure 5 shows the annual death rate from influenza in England and Wales. The disease, as will be seen, had its ups and downs for the next twenty or thirty years. Then in 1918 came the biggest and most lethal pandemic in history. This began in the summer, almost simultaneously around Boston and Brest, ports respectively for the embarkation and disembarkation of American troops coming to Europe. It very soon spread round the world and is believed to have caused between fifteen and twenty million deaths. In contrast to what has been recorded before or since, the worst of the mortality was among young adults, who died of a rapidly fatal kind of pneumonia. A feature of this was a lack of oxygen causing a remarkable colour in the face referred to as 'heliotrope cyanosis'. Some of the deaths were doubtless due to a pneumonia caused by the virus alone, but in most instances bacteria as well as

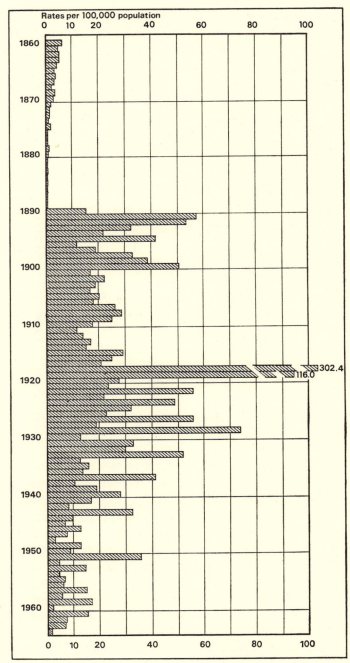

Figure 5. Death rate from influenza during the past century in the large towns of England and Wales.

virus were involved, streptococci, staphylococci, pneumococci and the
bacillus named after Pfeiffer and once believed to be the cause of influenza:
different bacteria predominated in different areas.

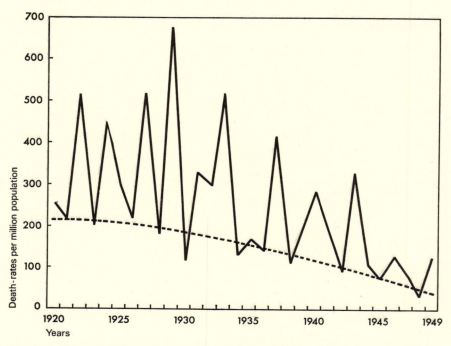

Figure 6. The graph shows a tendency, evident until 1949, for influenza to kill
fewer people.

As in 1889–90 there were several waves in this pandemic. After 1919
things gradually settled down to something more normal. In particular the
proportion of deaths in people under fifty-five fell from eighty-six per cent
in 1918 to forty-one per cent in 1933 and twelve per cent in 1951. In the
period from 1920 onwards there has been a tendency for flu outbreaks to
come every other year, but, as we shall see, there is hope that they may be
becoming less frequent. Figures 5 and 6 show that influenza is apparently
growing less troublesome. Not only are the peaks of annual mortality
tending to become lower, but the level of the troughs of the waves seems
to be falling too. We shall be returning to look at this diagram again later.

Knowledge about influenza began to have a more scientific and precise
basis in 1933 when an influenza virus was found to be able to produce a
feverish snuffling disease in ferrets. I was very fortunate in being con-

cerned in this early work: indeed the first virus to infect a ferret came from my own throat. A year later a way was found of infecting mice also; later still virus was grown in fertile hens' eggs and in tissue-culture and in 1940 the haemagglutination test was discovered. George Hirst [78] in New York noticed, when he was collecting fluids from flu-infected fertile eggs, that blood-cells accidentally reaching those fluids ran together into clumps. A whole battery of techniques now became available for the study of influenza and we probably know more about it than about almost any other virus. Fortunately we no longer have to observe and take daily temperatures of strictly isolated ferrets over a period of several days: instead we get results with the haemagglutination test within an hour or two. Ferrets, however, still have their uses in flu research; they are the best animals for producing the specific antibodies required in some aspects of research.

The haemagglutination test has been the tool beyond all others which has enabled us to sort out the differences between one flu strain and another. Using so far as possible standard amounts of reagents, one finds out what is the highest dilution of a serum which will stop haemagglutination by a particular virus – its titre. The figure shows the sort of result obtained by putting up tests with several flu viruses and corresponding antisera. It shows that the titre of a serum against its own virus has a high value, against related viruses the crossing is not so good in either direction and against a distantly related virus there may be little or no effect at all. By tests of this sort much has been found out about the relationships between different flu strains.

One of the first things to come to light was that there were viruses now known as influenza B which had nothing at all in common with the original (Influenza A) virus, by any serological test. We shall refer to their disease-producing behaviour later (see p. 66). Influenza B is a good deal less important than is A; influenza C, described later, is of smaller importance still. Then it appeared that the influenza A strains themselves were not all alike. They all have the same nucleoprotein antigen in the virus-core, as can be shown by the complement fixation test: but by neutralization and haemagglutination-inhibition tests they may differ. These are related to substances on the virus surface. The unexpected and important finding was that while viruses from one outbreak were pretty much alike, there was often a change between one epidemic and the next. In 1946 was born a suspicion that the changes in virus-behaviour would have to be studied on a world-wide basis: epidemics apparently moved across the face of the earth and changes in the influenza viruses in different countries seemed not to be changing in a separate and independent manner.

So, after discussions at an international conference in Copenhagen in 1947, the World Health Organization set up a world-wide network of laboratories based on a World Influenza Centre in London and a corresponding laboratory in the USA [5]. The plan was for laboratories concerned, in different countries, to isolate viruses from local outbreaks and send them to London or other central laboratories for comparison with other viruses. In turn, peripheral laboratories would be supplied with information, with samples of new viruses which might turn up and with laboratory reagents. One important object of all this was in relation to vaccine-manufacture. It is of no use for a manufacturer to prepare a lot of flu vaccine ready for the next outbreak, only to find that the virus has become changed in the interim and that his expensively produced vaccine may as well be poured down the drain. If, however, the WHO network can tell a European health department that a new variant of influenza (sample enclosed) has appeared in South Africa and seems to be spreading, that gives the local laboratories a chance to be ready with the right vaccine before the epidemic hits their country. That is the idea: in practice it is not quite as simple as is implied.

In the years since these international studies have been going on, certain things have become clear. First, there do seem to be successive changes taking place in influenza viruses all over the world, and the new strains seem commonly to spread round the world. The changes occur between one out-

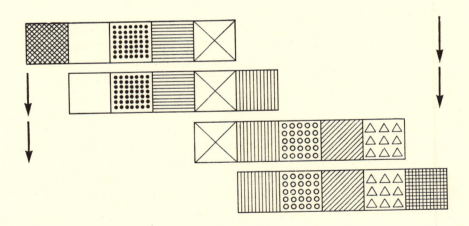

DIAGRAM OF ANTIGENIC DRIFT

Figure 7. Hypothetical figure showing changes in antigenic composition of influenza A in successive epidemics.

58

break and another, not in the middle of an outbreak. The changes are often of a minor character with new strains being fairly close to recent ones, rather more remote from those at a greater distance in time [6]. The process has been called 'antigenic drift'. A diagrammatic representation of the sort of thing believed to happen is shown in figure 7. This supposes that chemical antigens indicated by the differently shaded blocks are changing in relative proportions with perhaps old ones being lost and new ones emerging. At longer intervals of perhaps a decade, rather more drastic changes seem to happen, so that vaccines against older strains instead of being rather inefficient against older strains are found to be completely useless. That happened in 1946–7 and again in 1957. The pre-1946 strains are now classified as Ao, the 1947–56 ones as A1 and the post-1957 or Asian flu strains as A2. Not every antigenically novel strain succeeds in spreading widely: occasionally such a novelty turns up and makes us fear that a bad pandemic is in the offing, and then nothing more is heard of it. Perhaps it 'just hasn't got what it takes' to spread widely.

Another remarkable feature of the whole business is that when new variants of the A viruses appear, older ones seem quite quickly to disappear almost completely.

This behaviour of flu is best illustrated by describing the events taking place in three outbreaks.

In 1948 the World Influenza Centre was just getting organized. In the autumn of that year the disease broke out in the north of Sardinia. It was studied there by Professor Magrassi [105] who noted its appearance in a number of villages simultaneously. Moreover cases turned up at the same time in shepherds living in a state of virtual isolation amongst their flocks in the mountains. Shortly afterwards the disease appeared on the mainland of Italy and thence in 1949 apparently spread to Austria, Switzerland, France, Spain and Britain as indicated in the map (figure 8). Cases were recorded in Ireland, Denmark and Iceland, but by the time it had got that far the wave seemed to have spent its force and no big outbreaks developed in those countries. Viruses were received at the World Influenza Centre from all over western Europe and by serological tests were like each other, yet all slightly different from 1947 strains. In the following year similar viruses turned up across the Atlantic. All this looked like a matter of direct spread, yet the simultaneous appearances in many separate Sardinian villages also suggest an activation of virus previously seeded into that area.

Such may have happened also in the more complex outbreak of 1951. There had been local outbreaks in Sweden in June and July of 1950, then nothing more till October when an epidemic began in Sweden and

59

Denmark. This apparently spread to the east coast of England, to other parts of Scandinavia, Germany and the low countries. Viruses isolated were variants of the A1 viruses, distinguishable from those of immediately preceding years and were called the 'Scandinavian type'; the flu they caused was mild in character. A few weeks after the Scandinavian viruses had reached eastern England, a much more vicious influenza hit Belfast and Liverpool. In Liverpool in particular there was high mortality, as high as in the 1918–19

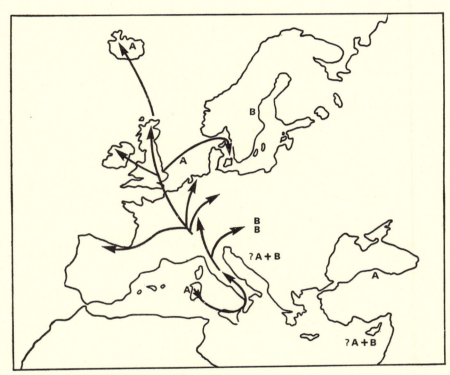

Figure 8. Spread of influenza A across Europe in 1948.

pandemic, but this time affecting the old and infirm rather than the young and healthy. The slightly different Liverpool type of virus turned up also in a number of countries around the Mediterranean and was found to be the same as strains isolated in Australia and South Africa a few months earlier (see figure 9). It looked, therefore, as if Britain and other European countries had sustained the impact of two waves of flu viruses, one coming from Scandinavia in the east, the other from the south. The two waves apparently met along a line drawn roughly from Dublin to northern England and so to the Netherlands and Italy. One cannot be sure that the high mortality in Liverpool was due to unusual virulence of the Liverpool strain; it may have been due rather to occurrence of much smog coinciding with the invasion by the flu virus. Moreover the Liverpool virus did not prove so fierce in other places.

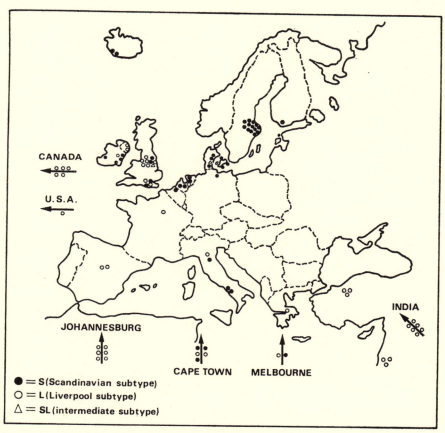

Figure 9. Distribution of two subtypes of influenza A in Europe in 1951.

The third outbreak I have to describe was a much more dramatic one, that of the Asian or A2 influenza in 1957. It apparently began in an area in Kweichow province in South China in February, spread over China in March and came to our attention when it reached Singapore and Hong Kong soon after. Chinese investigators quickly spotted that it was of a new antigenic type, but there was a two-months' delay before the rest of the world knew this: China was at that time almost the only country not

cooperating in the WHO influenza programme. The new virus was so different from its predecessors that populations had no effective immunity
against it and it therefore spread quickly right round the world. Figure 10
shows how the extent of the outbreaks waxed and waned in different
countries. In tropical countries, as the figures show, outbreaks started sud-

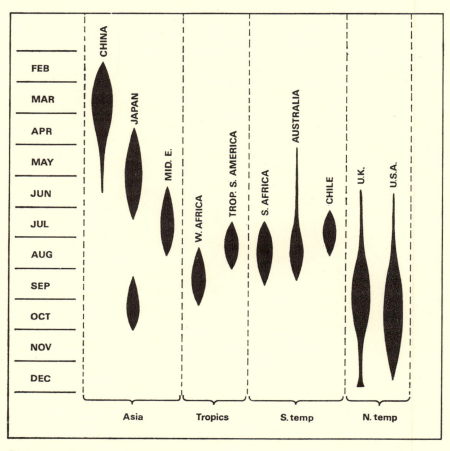

Figure 10. Waxing and waning of Flu.

denly, spread rapidly and only lasted for a couple of months or so [8]. Quite
different was their behaviour when they reached the USA, Britain and some
other European countries. There was evidence of introduction of the virus
during summer months and local outbreaks occurred, but these quite failed
to spread. Only when rather cooler autumnal weather came along were
there big outbreaks in temperate countries; when they did come they

62

definitely lasted longer than they had done in the tropics. It seemed that summer in temperate zones did not offer the right conditions to permit the Asian virus to get going properly. It is in that case very puzzling as to how it managed to do so without any difficulty in the tropics.

The Asian virus has presented us with quite a series of interesting problems. First, as to its origin. Other new flu strains are demonstrably related fairly closely to their immediate predecessors, less closely to those rather earlier. A2, however, differs quite widely from the later A1 viruses, not only antigenically but in quite a number of other biological properties. Could it have quite a different origin? Its first appearance in China is of interest. It may be remembered that the 1889 flu was believed to have started in Central Asia, and some still earlier outbreaks are rumoured to have had similar origins. There are influenza A viruses known to affect species other than man – pigs, horses and birds. It has been suggested that the Asian flu was derived from some Asiatic animal in which the virus is endemic, adaptation to man having taken place in the 1880s and perhaps earlier, and again in 1957. More definite evidence, unfortunately, is not forthcoming and the idea remains in the realm of speculation.

Another point of interest is that antibodies to the A2 virus were found by Professor Mulder of Leyden to be present in sera of some people aged seventy or eighty, people in fact who might well have had contact with the 1889 virus; from younger people such antibodies were absent [118]. It is suggested, therefore, that 1957 witnessed a return of the 1889 virus. There are two views as to the nature of the antigenic changes which take place in influenza viruses. According to one view there are a limited number of antigens in a flu virus; these are periodically shuffled, taking turns to be the dominant surface antigen which determines how the virus behaves in serological tests. Proponents of this view hold that the turn of the 1889 antigen had come round again. Alternatively it can be maintained that the virus's surface antigens change in their chemical make-up from one outbreak to another as a result of mutations and that all the possible components are not present all the time. Those holding such a view think that the apparent relationship between 1957 and 1889 viruses is due to a broadening of the range of reactivity of the old people's sera such as is known to occur after antigenic stimulation over a long period.

The last important thing to record about the A2 virus is that it seemed for a time at least to be antigenically much more stable than the A0 and A1 viruses. Instead of changing with each outbreak, it varied relatively little between 1957 and 1963; such minor changes as were detected were not big enough to be likely to influence the effectiveness of vaccines and were not on a world-wide scale. If this virus was indeed relatively stable one might hope that the immunity of peoples would be solid enough to keep the virus at bay better than in the past. Indeed in Britain and elsewhere, outbreaks have been lately showing an encouraging tendency to occur less frequently. Since 1963, however, variations in the A2 virus have been rather more marked.

We have some understanding of the mechanism by virtue of which flu viruses can change. When it is trying to circulate through an immune population there are naturally selective influences favouring any variant which is sufficiently different antigenically to stand a chance of overcoming the specific immunity it is up against. This can be demonstrated experimentally. If an influenza virus is passed in series through partly immune mice, viruses emerge which are antigenically different and so better able to survive. If mice are rendered immune by vaccination against several strains of virus current in previous years viruses have emerged having novel antigens. These experiments raised hopes that it might be possible, as it were, to synthesize ahead of time the variant which was likely to turn up and cause the next epidemic, but such has unfortunately not proved practicable. In 1955 a new type of virus turned up at just about the same time in India, Wales, Ireland and New York State. It did not cause much trouble but it raised an interesting question. Had there been in these countries simultaneous mutations in a similar direction by viruses facing similar hostile antibodies? Or had there been a widespread seeding of virus without disease-production so that it was ready and waiting to cause trouble when things were favourable for it?

Here we are again up against the question raised in the last chapter concerning the effect of season and weather on respiratory infections. The failure of the Asian flu to get under way during the European and North American summer seems very suggestive. We cannot at present explain why flu does not quickly start up with winter chills as colds do, waiting rather till the New Year. Possibly it does start up in the autumn but warms up slowly and then suddenly bursts out like a kettle boiling over.

It is worth looking more closely at Shope's work [147] on swine flu referred to briefly on p. 9. He was very puzzled by the sudden epidemics occurring simultaneously in many farms and he succeeded in obtaining evidence for the seeding of virus well ahead of the outbreaks. Apparently swine flu, which is a porcine representation of the influenza A virus family, is able to pass into the lung-worms which infest many pigs in the mid-west of America. It then passes out with the lung-worm ova in the pig faeces. These are taken up by earthworms in which the lung-worms pass several stages of their life-cycle. In due course the rootling pigs eat these earth-worms, and the lung-worms are able to pass, still bearing some swine-flu virus, into the pigs' lungs. Here for a time nothing happens, but the infection has been implanted and can be activated by a sufficient stress (see figure 11). In the ordinary course of nature it seems that the onset of cold wet weather is the stimulus which triggers things off. In the laboratory, Shope was able to produce the disease in his 'prepared' pigs by giving several injections of killed cultures of a bacterium – a *Haemophilus* related to

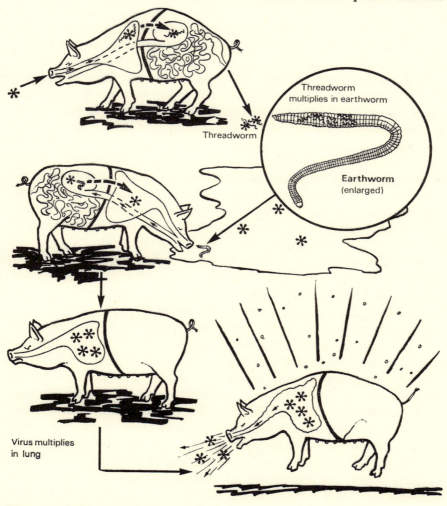

Figure 11. Supposed cycle of swine-influenza virus through lungworms, earthworms and back to the pig.

Pfeiffer's influenza bacillus or by injections of calcium chloride or by continuous spraying of the pigs with cold water. The swine-influenza virus produces only a very mild disease in pigs: clinical swine flu is seen when the *Haemophilus* just mentioned is present along with the virus. This is usually present in pig herds but does no harm unless the virus is there, too. Not everyone accepts Shope's explanation of the facts. It is unfortunately the case that the swine-flu virus cannot be directly demonstrated in the earthworms in his experiments; it is only revealed by the indirect test of feeding the earthworms to the pigs and then 'provoking' the infection by the stresses which Shope employed. There are, however, facts not readily explained except on Shope's ideas: pigs held in isolation in New Jersey, where swine flu is not naturally present, could be fed with worms from a swine-flu area in Iowa and then 'provoked' into developing the disease.

The influenza B virus, unlike A, is not known to affect naturally any species other than man; it is also less readily adapted to infect animals experimentally. It has caused widespread outbreaks as in 1946 and 1947 but these have usually been less extensive than the A outbreaks and have tended to come at more widely spaced intervals, perhaps because B seems to give rise to more solid immunity than A. B viruses do change antigenically, but less frequently and dramatically than A and their behaviour on a world-wide scale is not so obvious as with A. They behave more like endemic viruses and in fact are rather apt to cause purely local outbreaks in schools and service training centres as adenoviruses do.

Influenza C offers a bit of a mystery. Antibodies against it are usually widespread in the population, yet disease due to influenza C virus is uncommon. It seems to be a sort of camp-follower, for isolations of virus have frequently been made in the course of flu A outbreaks.

Many of our ideas about flu are based on study of charts of incidence such as are shown in figures 5 and 6. There is a difficulty in the interpretation of these since flu is not a notifiable disease nor one which can be diagnosed accurately on purely clinical grounds; and of course laboratory methods are only applied to a very small proportion of cases. We are now up against a very puzzling phenomenon. Figure 5 shows that influenza has caused a substantial number of deaths ever since 1889. Figure 6 shows the same and shows also that the numbers of deaths in non-epidemic years steadily fell up to 1948. But – and it is a very important 'but' – in these non-epidemic years it is very difficult by laboratory tests to show that either influenza A or B is present at all! What then is causing deaths in the non-epidemic years and what is it which has been slowly declining, as flu declined for nearly thirty years after 1920. Can there be flu about, even lethal flu, which laboratory tests do not detect? One fact makes us consider this as a possibility. Whenever there is a big flu outbreak, the virus is particularly apt to kill old people and those with chronic heart or lung disease and it also seems to kill many of these when their doctor does not realize that the virus is concerned, for influenza is not mentioned on the death certificate. The reason for saying this is that during a flu epidemic, deaths from all causes greatly increase. In fact when the Registrar-General's returns show a large increase in deaths from all causes, it is a safe bet that influenza virus will be at large and very active.

Despite all we have learnt in the past fifty years, influenza still presents us with many puzzles. We have to try to understand them for we do not know when another lethal pandemic may not strike us, as it did in 1918. It has been suggested that the 1957 Asian flu might have been as disastrous as 1918 flu if we had not had available the antibiotics capable of controlling the bacteria which certainly caused most of the 1918 deaths: that, however, is a matter of opinion. It seems that in spite of their triumphs, the existence of flu viruses is rather hazardous; or why were they in eclipse for several decades of the last century? If we understood more, perhaps we could eclipse them again.

Many of the puzzles concerning the epidemiology of flu might be resolved if we could establish what happens to flu viruses between epidemics. At the end of the chapter on colds I suggested that rhinoviruses and others might be able to pass from person to person at a low level of activity, only establishing a local foothold on mucous membranes and not engendering either disease or generalized immunity – a sort of tip-and-run affair. The same could be true of influenza and could afford a possible mechanism by which virus might be seeded into a population ahead of an evident outbreak. It would also fit in with the occasional occurrence, in a few more susceptible people, of antibody-rises or even clinical flu in between epidemics.

Chapter 7

Poliovirus and its Relations

THE VIRUSES CAUSING poliomyelitis together with the Coxsackie and Echo viruses are (see p. 15) included under the heading of 'enteroviruses' since their normal habitat is the alimentary canal. We have, then, to consider with these viruses dissemination of infection by the intestinal route, that is, through the medium of virus-infected faeces. The matter is, however, not quite as simple as that. The polio viruses, and perhaps the others, multiply especially in the pharynx and small intestine. These are both regions where a lot of lymphoid tissue lies beneath the epithelial cells lining the alimentary canal. Many people think that poliovirus multiplies in this lymphoid tissue, but no microscopical changes have been detected which can be ascribed to growth of the virus in this situation; so there is still dispute about the matter. In any case the result is that the viruses in question can be found in, and are doubtless shed from, pharyngeal secretions as well as by way of intestinal excreta. So their spread may involve both respiratory, and intestinal routes. All the enteroviruses are found very commonly in the intestines, especially of children, less and less commonly as age increases. They are, however, only there transiently and especially in late summer and early autumn months. They are much commoner in tropical countries and elsewhere where standards of hygiene are low; here the infections they cause are more uniformly spread through the year. The reasons why poliomyelitis, for instance, is so prevalent in late summer in temperate zones are just as obscure as the reasons, discussed in chapter 5, why colds are commoner in winter.

We shall consider in this chapter, first, the behaviour of naturally occurring poliomyelitis; then the bearing of its natural history on efforts to control it by vaccination; and finally relevant matters concerning the other enteroviruses.

The poliomyelitis viruses belong to the picornavirus family containing the very small riboviruses, and they are of three serological types which overlap very little in antigenic composition. They normally cause inapparent infections or very mild fevers; virus probably multiplies then only in pharynx and intestine, at most reaching regional lymph-nodes. Excep-

68

tionally it spills over into the blood-stream and may reach and infect the central nervous system, especially the anterior horn cells of the spinal cord. These may be temporarily or permanently damaged; in the latter event and if enough of them are destroyed permanent paralysis results. In classical poliomyelitis a first feverish bout is separated by a clear interval from a second bout in which the nervous damage occurs. This happens in a few cases, probably 1 in 100 with a virulent virus, only 1 in 10,000 with milder viruses.

We do not fully understand why a normally harmless virus can, in a few unlucky people, lead to devastating results. The genetic make-up of the individual probably plays a part. Age certainly does; older children and young adults, if they are infected, are more likely to suffer from paralysis than are small children. This is probably partly a question of naturally greater resistance in younger subjects. Also, very small children, in heavily infected areas, may receive a small dose of virus at a time when they are under the umbrella of the temporary protection of antibody passively received from their mothers. Then, again, fatigue may make paralysis more likely; if a person is infected at a time of great physical exertion, paralysis may affect the muscles most heavily involved. Injections of vaccines of various kinds, when given to a person already incubating the infection, may predispose to paralysis, and this is likely to be in the actual limb receiving the injection. So, if polio is prevalent, it is usually wise to postpone any inoculations which are not urgent. It turns out that when a case of polio is recognized in a household, it is almost certain that the other members of the household are already infected, usually inapparently, but with virus present in their stools. Any attempt to stop the spread of the disease by quarantine must take account of the fact that the case or few cases observed are the equivalent of the little fraction of the iceberg that shows above the surface of the ocean. A proviso is, however, needed: in some outbreaks due to an unusually virulent strain of virus, there is fairly clear evidence of person-to-person spread in what has been called a 'narrow stream of infection'.

An astonishing feature of poliomyelitis is the way in which it has changed from being a disease of early childhood, generally known in consequence as infantile paralysis, to one affecting older age-groups. In Sweden the change began about 1911. At that time only twelve per cent of patients who suffered paralytic disease were over twenty years old: after that the percentage steadily rose till it reached fifty-seven per cent in 1953 [121]. Similar figures are reported from other countries; only the times of the change are different. The explanation has gradually become clear. With improvements in hygiene, particularly with the general use of the water-closet, intestinal diseases such as typhoid and dysentery have become less and less of a menace. But, on the contrary, poliomyelitis has become worse! The bacilli causing typhoid and dysentery have been largely eliminated, but for the

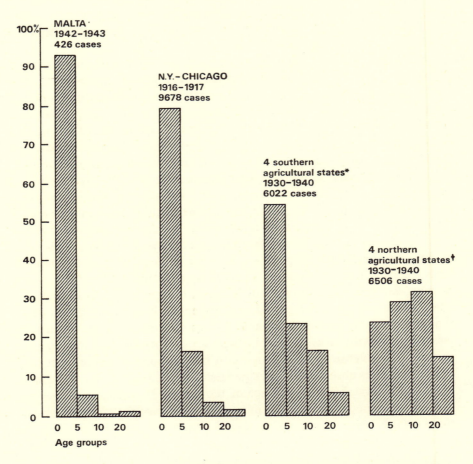

Figure 12. Numbers of cases of paralytic polio in four age groups, in areas with various standards of hygiene.

poliovirus virus hygiene has not achieved this. Where sanitary standards are low, babies are born and grow up in an environment where polioviruses of all three types are so prevalent that the infants are almost bound to be infected before they are many months old. So they get their infections at a time when their resistance is high and all are immune by the time they reach the age when an infection would have been likely to be a paralysing one. At first sight it seems anomalous that a disease should be called 'infantile paralysis' when infants are more resistant than older persons. But when hygiene is poor the disease, even if it only paralyses a very small percentage of the infants it attacks, will altogether paralyse an appreciable number, for almost all are infected; while it will do this to very few of the older ones because they will have been naturally immunized in infancy. Now, see what happens with improving hygiene: babies are progressively less certain of picking up the virus and getting an inapparent infection when they are small. They therefore grow up having no antibodies, or not against all three types; and when they do meet the virus there is a greater chance that the infection will be a serious one (see figure 12).

Polio travels with the crowd. In a country such as Sweden there was recorded a difference between urban and rural populations. In towns the virus spread more readily among children and there was more polio at an earlier age. In the country the virus did not get about so freely, infection occurred at all ages up to the mid-twenties and the total incidence of paralytic polio was more than twice as great for a population of a given size. There has been, however, a tendency for the difference between town and country to become less, even before vaccination changed the whole picture.

In isolated communities reached by the virus only after an interval of many years, all ages are likely to be affected. The extent of virus-spread among children can be gauged by estimations of the level of antibodies in their sera. In Cairo in 1952 almost all children were found to have acquired antibodies to all three types by the age of two, while in a city of comparable size in the USA few had done so [125]. In South Africa antibodies appear more generally and at an earlier age in Africans than in Europeans, doubtless because of lower levels of hygiene under tribal conditions. There is correspondingly a tendency for cases of paralytic polio to occur among them almost entirely under the age of five in contrast to incidence at older ages in Europeans. Now, where Africans are better housed and with better sanitation, polio is coming to behave in them according to the 'European pattern'.

There has been much argument as to the relative importance of pharyngeal secretions and faeces in the spread of the disease. It seems almost certain that under primitive conditions it is the faeces which are of major importance; otherwise why should improvements in hygiene have such tremendous effects? Even with good sanitation there is a notable decrease in the amount of spread of virus from small children as they pass from the pre-toilet-trained to the toilet-trained age. There is evidence that adults are particularly likely to pick up infection from a child, but we cannot stress that point seeing that the same applies to the common cold. With viruses of low virulence the amount in the pharynx is so much lower than in the intestinal excreta that spread by the latter is probably more important. With more virulent strains in communities with good hygienic standards the two routes may be equally important. It must be remembered, however, that virus is rarely present in the pharynx for much longer than a week, while it can often be recovered from the stools for two to four weeks, or longer. One can perhaps sum the matter up by saying that the person in countries with better hygiene acquires his infection later in life because he is in less danger of picking up the virus by the faecal route and is exposed only to the rather less efficiently spreading pharyngeal virus. After all, we were discussing in chapter 5 how colds were spread by sneezing and coughing; infection of the pharynx by poliovirus does not lead to any symptoms which would help to spread virus by such means.

Poliovirus has been frequently recovered, as one might expect, from sewage. There is, however, no evidence that this has ever been a source of infection. Water-borne infection such as may cause typhoid and dysentery seems not to occur. A few outbreaks of polio have, however, been traced to infected milk, always because a virus-carrier has been concerned in the distribution of unpasteurized milk. Polioviruses have been not infrequently recovered from flies such as blowflies and others which frequent filth. They may carry virus for two weeks and one investigator has claimed that limited virus-multiplication may occur in them. Chimpanzees have been infected by eating food contaminated by virus-carrying flies. There is, however, nothing to suggest that flies are of more than secondary importance, serving as an additional means of spreading virus around man's environment.

The ways of controlling virus diseases will be fully discussed in chapter 24. The question of polio-vaccination is, however, so closely tied up with knowledge of the natural history of the disease that it seems best to consider the matter now. As is generally known, there are two weapons available against polio; the killed vaccine named after Dr Jonas Salk and the attenuated live vaccine associated with the name of Dr Sabin, though a live vaccine was, in fact, introduced earlier by Dr Koprowski. Both types of vaccine are made from virus grown in tissue-cultures usually of monkey kidney. The killed vaccine has to be given in several doses with 'boosters' at a later date since killed vaccines do not in general give rise to permanent immunity. It has been urged that children with immunity following this vaccine are sure sooner or later to encounter wild polio viruses and that this will stimulate their immunity and make it permanent. There is, however, the possibility that the anti-polio campaign may be so successful that polioviruses will be eliminated from the environment, and then the hoped-for reinforcement of immunity will not occur. Most experimental evidence suggests that killed vaccines produce or increase the antibodies in the blood and protect against the danger of paralysis, but do not prevent establishment of viruses in the gut; they will therefore by themselves not eliminate the viruses. This view is not everywhere accepted. Vaccination with killed polioviruses in Sweden was successful beyond all expectation: after four years the disease had virtually disappeared. Most surprising was the fact that polioviruses were hardly being isolated at all and it was suggested that with a really super-excellent vaccine such as the Swedes claim to have, viruses were prevented from even growing in the gut. The Swedish results do not, however, altogether agree with those from the USA and elsewhere, so judgment must be reserved.

The live viruses used for vaccination have been attenuated by passage in rodents or in tissue-culture and by very careful selection. The consequences of using them on a massive scale such as have been employed in the USSR are of great interest. They are given by mouth, commonly on a lump of sugar or in a sweet and are often referred to as oral vaccines. There has been much argument as to the danger of reversion of the attenuated viruses to virulence, not necessarily in the person to whom the virus is given, but after passage to his contacts, perhaps in series from one to another. This reversion probably did take place with some of the viruses earlier used. With later ones some reversion occurs in the shape of partial reacquisition of ability to cause disease in monkeys, but the evidence suggests that the change is not great enough to be of serious danger to man. Obviously the danger would be less if virus were fed simultaneously to whole communities, so that all received the attenuated virus together and there were no susceptibles available for serial passage of the virus. It is a very difficult matter, in a field trial, to determine whether any reversion to virulence has occurred. If, after widespread use of vaccine, an odd case or two of polio turns up, who is to say whether it arose from the live vaccine or would have happened anyway as a result of picking up a 'wild' virus? Fortunately there are ways of settling the question. Viruses are used for vaccination having 'markers', unusual characters of one sort or another, mostly concerned with their behaviour in tissue-culture. If cases of polio do turn up in a vaccinated community, one can therefore usually say, by noting the time of its occurrence and the presence or absence of the characteristic markers of the vaccine strain, whether it was a wild virus or came from the vaccine. In almost every instance it has been possible to pronounce the vaccine as innocent.

An important question concerns the freedom with which vaccine strains spread. Certainly they do so very well within a household and in institutions where children mix freely together. Spread from household to household is not so good, but it is better when hygienic conditions are imperfect. Vaccine strains seem, however, to be less efficient spreaders than are wild polioviruses. One might hope for very good spreading with the idea that seeding vaccine-virus to only a proportion of people might immunize all the susceptibles. This has raised the ethical question of whether people ought to be immunized against their wills. There have even been questions raised about the spread of vaccine virus across frontiers from a country which believes in live vaccine to one which does not. These arguments are in fact rather academic ones. What will possibly happen is that if the live vaccines are used on a big enough scale all over the world they may drive out and replace the virulent viruses; let us hope so.

Live vaccines offer more prospect than do killed ones of the development of a permanent immunity such as follows natural infection with measles and mumps. Long, perhaps life-long, immunity results from giving yellow-fever vaccine. On the other hand vaccination against smallpox, though a live vaccine is used, only endures for a few years. It is too early to say how durable will be the immunity which follows the use of live polio-vaccine. It seems, however, that these live viruses do protect against alimentary infection as killed vaccines probably do not.

A complication arises in attempts to protect with live vaccines and this concerns the existence of the three types. All three must gain temporary establishment in the gut if they are to induce immunity; but each of the three is liable to interfere with the other two, by the mechanism discussed in chapter 4. It is obviously administratively easier to give the three in a combined operation and this is often done. Some think it better, to avoid risk of interference, to give them in turn. Type I is then given first as it is the one most commonly causing paralysis and therefore most urgent to protect against, then some days later type III as of next greatest importance and finally type II. According to different circumstances and in different countries, the types are given one, two or three at a time.

Interference may take place, however, not only between the three polio viruses but between them and other echoviruses and even adenoviruses. This is a particularly serious matter in tropical countries where thirty-two echoviruses, thirty coxsackies and twenty-eight adenoviruses may be in circulation and any child is liable to have one or other of the ninety in his intestines at any one time. Oral vaccines in such conditions are liable to give less than perfect immunization at the first attempt – in one trial there was only a thirty per cent 'take' of polioviruses given by mouth. The difficulty can be overcome by giving repeated doses or by choosing a time for administering them when the interfering viruses are less likely to be prevalent.

One other trouble should be mentioned. When live polio virus is given there is a danger of its being contaminated with one or other of the numerous 'simian viruses' which are liable to be present in apparently normal monkey kidneys. Presence of one of these was for a long time overlooked as it produced no changes in cultures of the rhesus monkey kidneys which harboured it. Its presence was detected only when transfers were made to cultures of kidneys of African monkeys: in those it did produce changes. Now that its presence can be recognized it can be eliminated; this is especially desirable as this particular virus, known as vacuolating virus or SV 40, can give rise to cancers in hamsters as will be described later (see p. 191). This virus has been present also in some widely used batches of killed poliovirus, as it is less readily killed than polioviruses themselves: fortunately there is no evidence that it has done any harm to the people who received it. SV 40 is a papovavirus (see pp. 18–19); others of the simian viruses belong to the myxovirus, picornavirus and adenovirus families and one is a herpesvirus. So the kidneys of monkeys are clearly a happy hunting-ground for anyone who is on the lookout for latent viruses.

The epidemiology of the Coxsackie and echoviruses affords extremely close parallels with that of poliomyelitis. All these viruses can be found as transient inhabitants of the alimentary canals in children, especially in late summer and autumn and most abundantly where hygiene is poor. Entero-viruses, more than ninety per cent of them echoviruses, occurred in sixteen per cent of children in Mexico City, while a corresponding figure for Cincinnati was five per cent. Moreover the incidence in the USA was three to six times as high in the lower as in the higher socio-economic groups [132]. Evidence suggests that most of these viruses are spread more by means of faeces than from pharyngeal secretions.

The characteristic feature of the Coxsackie viruses, from the laboratory point of view, is their ability to produce fatal infections in suckling mice. They are divided into two groups, A and B, according to the type of illness caused in the mice [170]. These two groups correspond to differences in the illnesses caused in man. The A's, with twenty-four serological types, cause a particular type of sore throat, called herpangina, sometimes meningitis and with one type, A7, even paralysis. Colds caused by Coxsackie A21 have been mentioned earlier (see p. 49). The B's, with six antigenic types, can give rise to illnesses with severe muscular pains in the chest and abdomen, as well as to meningitis. In South Africa, and later elsewhere, Coxsackie B viruses were found to be causing fatal heart-disease – myocarditis – in new-born babies; several outbreaks occurred in maternity homes [61]. It is suggested that in areas with better hygiene, there may be mothers who have grown up without encountering these Coxsackie viruses; if in a maternity home the virus is introduced and reaches the babies, these will have no antibodies from their mothers and will therefore be very vulnerable. There seems to be no natural resistance of very young babies to Coxsackie B.

The echoviruses grow more readily than the Coxsackies in tissue-cultures, especially of monkey kidney and do not normally infect baby mice. They also can cause meningitis or fever with a transient rubella-like rash, especially in children. As with polio, if adults do develop meningitis, it is likely to be more severe than in children. Normally the echo and Coxsackie viruses, again resembling polio, cause only inapparent infections; only the unlucky few become ill. At times, however, strains turn up with greater potentialities for trouble. In 1956, Echo 9 began to cause fever and rashes, also meningitis, in South Africa and Italy. Soon after there were outbreaks in Britain and over much of western Europe and later the same happened in Iceland and North America. One doctor in the north of England estimated that three to five per cent of all the patients in his practice must have been infected within a short period. The proportion of obvious to silent cases seems to have varied; also the relative numbers in adults and children. The properties of this virus illustrate the difficulties in virus classification. The originally described strains of Echo 9 behaved in a conventional manner. In contrast, many of the epidemic strains infected suckling mice as Coxsackie viruses do, and some antigenic relationship was shown to one of the Coxsackie viruses. In other respects, however, it still behaved as an echo virus should. All of this just shows that viruses hate to be regulated and placed in tidy compartments, preferring to present the taxonomist with awkward puzzles.

The ecology is broadly similar for the related viruses affecting other species. In Czechoslovakia and in Madagascar there are enteroviruses of pigs causing very serious outbreaks of a paralytic disease, known, from one locality where it was troublesome, as Teschen disease. In Britain and other countries antigenically related viruses occur, but the disease is often inapparent and occurrence of symptoms due to nervous system damage is only sporadic. So far as is known, the differences are those concerned with the virulence of the strains of virus prevalent in these different parts of the world.

Chapter 8

Mosquitoes, Horses, Birds and Snakes

IN COMING TO THE arboviruses we enter the most fascinating and complicated field of virus natural history. As previously mentioned, 'arbovirus' is a telescoped form of arthropod-borne virus. Of about 170 known viruses in this category, the large majority are carried by mosquitoes, and some of these we shall deal with in the next four chapters, leaving the tick-borne ones to chapter 11.

Very many of the arboviruses have been recovered only from mosquitoes, ticks or other arthropod vectors or else from rodents or other species undergoing inapparent infections. For most of these no relation to disease is known. Some regions of the tropics are particularly rich in arboviruses and even within one such region certain quite small localities yield far more viruses than do other places not obviously so very different. Parts of the Amazon basin, swamps in Trinidad and parts of Togoland are examples of such areas. Isolations have been made particularly by the aid of what are called 'sentinel' animals. These are susceptible animals in cages or tethered out in the bush so that they may be bitten by infected vectors and so pick up virus. In early days with this technique, monkeys were used. These required a lot of attention; now that mice, particularly very young mice, have been found to be very susceptible to many arboviruses, these are largely used instead. They can be put out in fairly open cages to which mosquitoes have ready access, or placed within traps designed to catch any mosquitoes which enter. An additional refinement is the use of a timing device designed to make catches only during certain hours and so reveal during what period of day or night the mosquitoes have made their attack.

Arboviruses carried by mosquitoes multiply, so far as they have been tested, in insects, usually those belonging to or closely related to the genera *Culex* and *Aëdes*. They multiply in the cells of the alimentary tract, then find their way into the blood-containing body cavity and so to the salivary glands, where they remain, all ready to be injected with the saliva when the female mosquito takes a blood-meal from a vertebrate: only female mosquitoes bite. This development takes a little time: so the mosquitoes are not normally infectious till the virus has undergone a cycle of development within them; this period is called the 'extrinsic incubation period' in contrast to the 'intrinsic incubation period' between the bite of the mosquito and the development of disease in the bitten man or other animal. Arboviruses, so far as is known, do no harm to their insect vectors; this fact suggests that they have probably arisen in insects who have made use of vertebrates to transfer the viruses to more insects. If they were originally vertebrate viruses using the insects in a similar way, we should not expect the insects to be so free from unpleasant consequences of virus-growth within them.

Most arboviruses are carried by only one species or a few closely related species of mosquito. In other mosquitoes the virus may multiply in their guts, but fail to pass into the body cavity, being held up by some barrier; so the virus never reaches the salivary glands and the mosquitoes therefore do not act as vectors. If, however, virus is artificially injected into the body cavity of such an insect, it will multiply. In fact some arboviruses have been shown to multiply when so injected even in houseflies, grasshoppers, bedbugs, beetles and moths. Such multiplication is, of course, of no importance in nature. Its occurrence is not very surprising; if a virus can multiply in tissues as different as those of a mosquito and a man, it should hardly be fussy enough to boggle at a diet of bed-bug.

In the infected animal, the virus of course multiplies again, not always in the same tissues; but it spills over into the blood, and viraemia, or presence of virus in the blood, is obviously necessary if virus is to be available to be taken up by another mosquito. This compulsory multiplication of the virus in two hosts means that its survival depends on the existence of a suitable relationship between those hosts. Not only that, but other vertebrates than the normal or 'maintenance' host come into the picture and so do other potential insect vectors. The various interlocking factors are so numerous and variable that it needs a complicated mathematical formula to determine what is likely to be the outcome, whether extinction of the disease, or endemic or chronic balanced infection or a spreading epidemic. It can roughly be said that if one successful host-to-host transmission regularly occurs we have an endemic situation, if less than one the infection dies out and if more than one an epidemic develops.

We shall have to consider in more detail some of the following factors: how long the vertebrate host and mosquito respectively are infectious one for the other; the amount of virus present in each and transferred to the other; numbers of vertebrate hosts and of vectors present in the environment; climatic factors; behaviour of the insects and of the vertebrates; interference with the environment by man; and factors concerned with immunity: a formidable list.

Mosquitoes once infected are usually infectious for life, but of course they do not feed very frequently. Infected vertebrates have viraemia for only a few days; if this is brief they may well escape being bitten during the infectious period. If an abnormal host is bitten, there may be so little virus in the blood that it is unlikely to be picked up by a mosquito or to be able to multiply if it does get in. This can be illustrated by the story of equine encephalomyelitis or encephalitis in North America. This inflammation of the brain (encephalitis) and spinal cord (myelitis) is sometimes popularly called sleeping sickness, but is of course quite different from the sleeping sickness of tropical Africa which is caused by a trypanosome, a protozoan parasite. Outbreaks of the disease in horses have been known for eighty years and in 1938, 184,000 horses in the USA were infected, with a very high mortality, sometimes up to ninety per cent. A similar disease occurs in man, and over three thousand people were attacked in 1941: incidence has always been lower than in horses. People affected may have fever and a variety of symptoms due to involvement of the brain – drowsiness, incoordination, convulsions and even coma and death. Horses show fever, often excitement, then stupor and paralysis. Two viruses have been concerned, known as western and eastern equine encephalitis, or in laboratory jargon, WEE and EEE. Their range is over the eastern and western parts respectively of the USA and southern Canada, with some overlap: WEE extends more widely. Both viruses extend their range to South America. The western form of the virus may infect people without causing serious illness. In endemic areas eleven per cent of adults are found to have developed antibodies. The eastern virus causes more severe illness with seventy-five per cent mortality and many serious sequels to the infection. Inapparent infections are rarer. The epidemiology of the disease was not understood until it was realized that both viruses naturally infect wild birds, perhaps especially water-haunting birds, and in these there is usually a wholly inapparent infection but with plenty of virus circulating in the blood. Eighty-six different species of birds are known to be susceptible either to the equine encephalitis or St Louis viruses (see p. 92). However, only two species, the house (English) sparrow and pheasant develop fatal infections in nature, and both of these are introduced species in North America. Certain mosquitoes bite both birds and mammals, including horses and man: and these carry the infection across. But in neither horse nor man is there much viraemia despite development of severe illness. So it seems unlikely that either species plays a part in a chain of infection: both are blind-alley infections, useless from the point of view of the virus.

A third kind of equine encephalitis, the Venezuelan, occurs in the northern half of South America and the Caribbean and has lately spread to the southern USA: the normal, maintenance hosts are probably small mammals rather than birds.

For an arbovirus to succeed in the world there must be reasonably high numbers within a limited area both of vertebrate hosts and of vectors. Virus survival probably depends on its circulation between a regular maintenance host or a small number of hosts and a regular vector. If this vector specializes in biting the hosts in question and does not indulge in too varied a diet, this will be all to the virus's good: little of it will be directed into the blind alleys referred to above. The abundance of vectors depends on climatic factors. If weather is too cold, mosquitoes are inactive. The relatively low temperatures of the north temperate zone have apparently limited the northward spread of mosquito-borne viruses; and the teeming millions of mosquitoes of the northern parts of Siberia and Canada do not seem to be concerned in transmitting arboviruses. On the other hand abnormally wet weather may lead to flooding and the provision of excellent conditions for abnormal multiplication of mosquito larvae. Outbreaks of virus-disease have followed the occurrence of plagues of mosquitoes after such flooding. What seems to happen is that virus unobtrusively circulates between birds and mosquitoes in a normal year, its presence being quite undetected because horses and man are not infected. When conditions change in the mosquitoes' favour, the incidence in the normal hosts rises and there is a spill-over with involvement of other vertebrates and other vectors. Eastern equine encephalitis affords an example. An outbreak involving horses and man occurred in New Jersey in 1959. Following an unusually high rainfall there was a great increase in the numbers of three mosquito species, *Culiseta melanura*, which is probably the main vector, *Aëdes sollicitans* and *A. vexans*. Subsequent events have been explained as follows [75]. A consequent spill-over of virus into mammals was helped by a change in the wind from southwest to south, bringing a number of salt-marsh mosquitoes into populated areas. *C. melanura* is likely to have introduced virus into those areas, infecting particularly chicks. Then *A. sollicitans* took over and transferred infection from chicks to man while *A. vexans* carried virus from infected chicks to horses. Another example of the effects of weather changes will be mentioned in the next chapter in relation to Murray Valley encephalitis.

The behaviour of the mosquitoes also affects the issue. Some populations of a common mosquito such as *Culex pipiens* bite only birds; other populations of the same species, perhaps living not far away, will bite man also. Much the same differences occur in other species. Some mosquitoes bite by day, others by night: nocturnal animals are likely to be bitten by day-flying mosquitoes for they are more defenceless when asleep, while the diurnal vertebrates will for the same reason be more apt to be victims of nocturnal mosquitoes. It is of interest that in parts of Utah, the local *Aëdes* bite so viciously in the evening that people are driven indoors; they thus avoid being bitten by the *C. tarsalis* which comes out later in the evening and is the carrier of the western equine encephalitis virus. In tropical jungles the insects tend to inhabit particular layers of the forest, and if they do this strictly, the mosquitoes of the forest canopy are likely to infect only monkeys and other species living aloft and not those dwelling at ground level. The 'vertical stratification' of mosquitoes and other biting insects has received a good deal of study, especially at Entebbe, in Uganda. Towers have been erected in the rain-forest with platforms at different levels (plate 21). Mosquitoes have been collected either in traps or by using human 'bait': – the 'bait' collects in tubes the hungry mosquitoes which come along. Much information is being gathered in this way, and there have even been papers on the relative attractiveness of 'washed' and 'unwashed bait'.

Where man intrudes by himself or with his domestic animals into a strange environment he of course exposes himself and them to the risk of being infected by viruses which are circulating harmlessly in their normal hosts. Danger to horses in the American continent has already been mentioned. In Africa the risks are greater and more complex and particularly in South Africa, where there is more raising of livestock and where this goes on in districts not far from where there are abundant antelopes and other wild ungulates. The natural hosts of many of the diseases of stock are unknown. There are not only blue-tongue in sheep and African horse sickness, both carried by *Culicoides* midges, but viruses other than arboviruses may come from wild animals, malignant catarrh, a herpesvirus, from wildebeest and African swine fever from wart hogs.

Clearing the jungle for purposes of agriculture alters the balance of nature, perhaps allowing excessive multiplication of hosts in which a virus flourishes particularly well or of dangerous vectors. Destruction of game is believed in one instance to have led to an increase in man-biting by certain mosquitoes deprived of their normal food. Extensive irrigation schemes have had unfortunate consequences as regards the risk of virus disease. In southern California such irrigation has favoured both *Culex tarsalis* and water-birds which are common virus-hosts and has created a virus problem which was not there before. Should it ever become possible to irrigate and fertilize the Sahara and other desert places, it will be important to guard against the possibility of similar happenings.

A factor which may limit the extent of a virus infection is the induction of immunity. As we saw in chapter 4, by spreading too successfully and immunizing whole populations of hosts, a virus defeats its own ends. As we shall see in chapter 10, something like this may happen with yellow fever (p. 106). There is probably an optimum dispersal of hosts and vectors in nature for the maintenance of an endemic virus infection. Some arboviruses probably do well because they can thrive in non-immune nestling birds present at the same time as an abundance of mosquitoes. In this and other cases the rate of multiplication of birds and mammals and consequent provision of fresh susceptible hosts is all-important.

In discussing influenza we were up against the difficulty of explaining how virus persisted between epidemics. A similar difficulty meets us with the arboviruses of temperate countries. How do they survive through the winter when mosquitoes are not actively spreading them around? There has been a host of fascinating theories put forward as to how overwintering is ensured, especially in relation to the equine encephalitis viruses; and we do not even yet know what is the truth of the matter.

First theory: virus might persist in mosquitoes from one generation to the next through ovum, larva and pupa without having to pass through a vertebrate – so-called transtadial (through stage) or transovarial transmission. This works all right for ticks, as we shall see (p. 109) but seems very unlikely in the case of mosquitoes.

Second theory: the infection dies out in northern latitudes in the winter but is reintroduced each spring by migrant birds returning to their breeding haunts. A great deal of work in collaboration with ornithologists has been devoted to this subject and a report on the possibilities was published by the World Health Organization in 1959. The viraemia in infected birds only lasts for a few days. One might therefore have to suppose that the bird picking up virus when bitten by a mosquito in its southern winter haunt would fly all the way, perhaps to Canada, with virus still circulating in its blood. This is thought to be rather improbable. It could be, however, that the bird was refuelled as it were at a stopping-place *en route*, with a mosquito transferring virus from an infected migrant to a non-infected one. At a station in Alabama on a migration-route, blood samples were taken from 242 birds passing through and not normally resident in the area. WEE virus was recovered from eight and EEE from seventeen; nine per cent in all. On two occasions the eastern equine virus has been isolated in Trinidad at times coinciding with waves of migrating birds. An argument against a seasonal transfer between North and South America is that there are definite though minor differences between the EEE viruses from the two continents [37]. Strains from Trinidad have resembled the South American viruses, while those from Jamaica have been like those from the USA. Consideration has to be given also to a possible role of migrating bats.

This whole question has been raised on several occasions with regard to the arboviruses of the Old World. Cases of acute encephalitis of unknown cause crop up from time to time in Britain: they could well be caused by arboviruses. Many migrant birds breeding in Britain spend the months of the northern winter in West and South Africa where there are numerous arboviruses liable to infect birds. A long-distance migrant bird could presumably carry such a virus from Africa to Britain as readily as from South to North America. As yet, however, there is no evidence at all for the occurrence of such virus-transport in the Old World: in fact Britain is, so far as we know, quite free from mosquito-borne viruses. Some search has been made for them in England, especially in *Aëdes* mosquitoes, but with completely negative results.

A newly discovered virus in the south-eastern USA has been christened 'Tensaw'; it is carried by *Anopheles* mosquitoes, seems to be fairly common, and infects rabbits, some rodents and dogs, but not birds. Transport by migrating birds is thus out of court as a mechanism for over-wintering; discovery of how the Tensaw virus survives may throw light on the problem as it affects the equine encephalitis viruses.

Third theory: virus can sometimes persist longer than is normally thought in vertebrate hosts. There is a record of persistence of the western equine virus in a bird for 234 days, though that was something very unusual. It is also possible that viraemia may recur in birds infected some time earlier under conditions of stress – a theory which may recall a similar suggestion (see p. 51) in relation to respiratory infections. Long persistence of virus in hibernating bats can occur and there is a likelihood that during hibernation it would remain at a low level, to be followed by considerable viraemia when hibernation was over in the spring. A new idea was born a few years ago concerning garter-snakes (*Thamnophis*) of a common harmless North American species, which may be bitten by mosquitoes though not very frequently or readily [167]. Viraemia in these snakes tends to be rather chronic, perhaps because their temperatures are lower than those of birds and mammals. It has been shown that both the eastern and western viruses will persist through the winter in garter-snakes kept out of doors; also (for EEE) in turtles kept through winter months in a refrigerator.

When snakes awake in the spring, there is enough viraemia to permit transmission of infection to chicks by mosquito-bite.

Fourth theory: virus overwinters in an invertebrate other than mosquitoes. Some mosquito-borne viruses have been shown to be transmitted by and to persist in ticks, in which, as we shall see, overwintering is not such a difficult problem. Epidemiological evidence is, however, all against the importance of ticks in the case of the equine viruses. Some years ago it was thought that the problem of virus-persistence had been solved by bringing bird-mites (*Dermanyssus*) into the story. Virus was recovered from mites biting nestling birds and in these, it was thought, virus might well carry over through the winter. Subsequent work did not support these claims and the mite-theory is now unpopular.

The fifth theory has the support of Dr W. C. Reeves [133], who has studied the subject more than anyone else; at least with regard to the western equine virus. The vector concerned is *Culex tarsalis* and this mosquito hibernates as an adult. Virus has been recovered from hibernating mosquitoes in nature from all except two months in the middle of winter. It appears that during summer months these mosquitoes prefer to bite species such as mourning doves and chickens around farm-yards, but as winter comes they turn their attention to small passerines such as sparrows, They do wake up in warm periods in the winter and take a blood-meal. Reeves thinks that the winter-cycle is probably much the same as that in summer, but that the whole thing is slowed down and progresses at an altogether lower level of activity both in birds and mosquitoes. There is, however, a suggestion that mosquitoes may not live so long after a blood-meal as when starved; and other workers have grave doubts as to the possibility that overwintered mosquitoes can achieve much in the way of spreading virus; they are few and bite very little compared with the following brood.

A sixth theory, only suggested quite recently [63], is based on the recovery of the Eastern virus from *Peromyscus* and other rodents during winter months. It could possibly be transmitted among them during the winter by some ectoparasite, only spilling over into birds when spring arrives. There would thus be a basic cycle in rodents, a secondary, summer cycle in birds and a spill-over from that to horses and man. The matter is, however, still unsettled. Possibly several methods are available for getting the virus over the winter months.

The eastern, western and Venezuelan viruses which have served to illustrate a number of points in virus natural history in this chapter belong to the A group of arboviruses (see p. 16). The eighteen or more members of this group all show some degree of antigenic relation and are alike in a number of other properties. Very possibly they are derived from a common parental stock. Two other A viruses now deserve our attention, each of them having been the cause in recent years of extensive epidemics in East and Central Africa: they rejoice in the names of Chikungunya and O'nyong-nyong. Most arboviruses have been named from the localities where they were first isolated and these names, including Bussaquara, Spondweni, Wesselsbron, Sathuperi and Bunyamwera, are quite sufficiently tongue-twisting for an Anglo-Saxon. Chikungunya and O'nyong-nyong differ in not being place-names but names in local dialects meaning twisting or bending-up, from the contorted position of the sufferer; for joint-pains are extremely severe.

The Chikungunya virus was isolated from patients in an epidemic occurring in southern Tanganyika in 1952 [136]. Besides the joint-pains there were fever and rash, but the disease was not fatal; the joint-pains and stiffness were apt to recur during a period of several months. The virus has been isolated since from Central and South Africa. The symptoms resemble those of dengue (see p. 93) and dengue has been thought to occur in Africa; but so-called African 'dengue' may have been due to Chikungunya and other arboviruses. How far Chikungunya in Tanganyika was a 'new' disease we do not know. It had probably not been widely prevalent for some time for the incidence in some areas was from forty to eighty per cent. It may be significant that in some areas people over forty-five years old are said to have been relatively resistant. The yellow-fever mosquito, *Aëdes aegypti*, seems to have been the main vector, though virus was isolated from two other mosquito species. Chikungunya virus has also turned up in Thailand; its behaviour there is dealt with in chapter 10.

Whether or not Chikungunya was a new disease, O'nyong-nyong carries all the hall-marks of being something novel [148]. In 1959–60 it swept through large tracts of Africa beginning in northern Uganda (see figure 13). Symptoms were similar to those of Chikungunya with the difference that lymph glands were often inflamed and swollen. It is estimated to have affected five million people, being thus the biggest arbovirus epidemic on record. The large majority of mosquito-borne arboviruses are carried by insects related to the genera *Culex* and *Aëdes*, but evidence is strong that O'nyong-nyong was transmitted by the malaria mosquitoes *Anopheles*, especially *A. gambiae* and *A. funestus*. Evidence has only recently been turning up that *Anopheles* mosquitoes must be considered more seriously in relation to arbovirus infections: transmission of Tensaw virus by an *Anopheles* was referred to on p. 83. O'nyong-nyong is more closely related to

Chikungunya than to other A-group arboviruses; antigenic analysis shows that it possesses at least one antigen which is not present in its relative. It is also related to another African A arbovirus, Semliki forest virus, which, however, is not known to cause any disease, though antibodies to it do occur in some human sera in Africa: *Aëdes aegypti* can carry it, as experi-

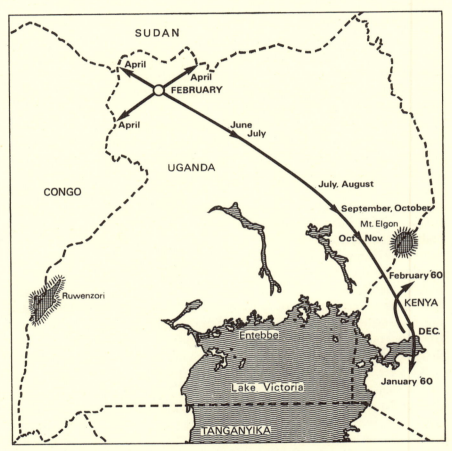

Figure 13. Spread of O'nyong-nyong virus through Uganda 1959–60.

mentally, an *Anopheles* has been shown to do. It has been suggested that O'nyong-nyong has been derived recently either from Semliki forest or Chikungunya viruses, and possibly its adaptation to grow in *Anopheles* rather than *Aëdes* mosquitoes was associated with a change in antigenic or other properties; these may have permitted it to spread in human populations in such a way as to cause the explosive epidemic we have seen.

Chapter 9

Mosquitoes, Herons and Pigs

THE VIRUSES BELONGING to the B group of arboviruses are related to each
other by antigenic tests much as the A-viruses are, though perhaps not
quite so closely. There are more than thirty of them and some are so important that they will claim most of our attention in the next four chapters.
Their behaviour illustrates, in a number of rather different ways, some of
the general principles discussed in the last chapter. Within the B-arbovirus
group are a number of subgroups and one of these subgroups contains
members rather closely related to each other by serological tests and probably owning a common origin. They seem to be naturally infections of birds
transmitted by *Culex* mosquitoes, infecting man and other species rather
incidentally, as do the equine encephalitis viruses we have just discussed.
They involve most of the countries in a great arc surrounding the Indian
and Pacific oceans. The West Nile virus is found in much of Africa, including Egypt, the eastern Mediterranean and round to India, where Japanese
encephalitis virus takes over and covers all the Far East round to eastern
Siberia. Murray Valley fever virus is an offshoot involving New Guinea
and Australia, and the St Louis encephalitis virus extends over the western
half of the United States, just reaching across to the Atlantic area to affect
some islands in the Caribbean and even Brazil. Not all strains are serologically identical and there may be strains intermediate between some of the
viruses. The Murray Valley virus, for instance, does not differ much more
from classical Japanese virus than do some Japanese virus variants from
each other.

Let us start in Africa and work round to the East. The West Nile virus
occurs in South, Central, East and North-east Africa, less certainly in the
west. It has been most studied in Egypt, which is perhaps its headquarters.
It is endemic in birds, being transmitted, at least in Egypt, chiefly by *Culex
univittatus*. Antibodies were found in eighty per cent of hooded crows but
also in very many house sparrows, herons and pigeons. Possibly *C. pipiens*,
which hibernates as an adult, plays a role in carrying it over the winter. The
infection, sometimes called West Nile fever, is endemic in Egypt, affecting
chiefly children, in whom the infection is usually silent. The infection turns

up intermittently in Israel where it causes outbreaks of epidemic disease attacking people of all ages [18]. Affected people have fever, sore throat, enlarged lymph glands and a rash rather like German measles. Meningitis occurs, but uncommonly. There have been suggestions that some viruses will destroy cancer cells more readily than normal cells; so West Nile among other viruses has been given to inoperable cancer patients in hope of benefit. Indeed some improvement has occurred, though only temporarily. However, eleven per cent of those inoculated developed symptoms of mild encephalitis [160]. Encephalitis due to this virus is recorded in horses, and perhaps donkeys, in Egypt. On the whole, however, the West Nile virus is less neurotropic, that is with less affinity for the nervous system, than the related viruses we have to consider next.

The antigenic make-up of the virus seems to be simpler than that of its relatives; for this and other reasons it has been suggested that it may represent the ancestor from which other viruses of the B group have evolved. The tick-borne viruses (p. 108) have also a claim to this position.

The natural history of the Japanese encephalitis virus affords us one of the most fascinating stories we have to deal with. It has been worked out by a team of investigators of the US Army, headed by Drs Scherer and Buescher and published in a series of papers in 1959 [141]. Their investigations were centred on the plain round Tokyo. They already knew when they began that the reservoir of infection was probably in birds. They paid particular attention to birds of the heron family, particularly black-crowned night-herons and egrets. These bred in two big heronries near Tokyo with several thousand birds in each (plate 22). They were convenient to study for several reasons. Many nests could be found together; they were accessible, many of them being in bamboo thickets and only six to eight feet from the ground; and the young birds were big enough to stand the taking of blood-samples. It turned out that a fortunate choice had been made; these herons did indeed prove to be the most important birds involved, partly because their breeding-season coincided with the time of maximum prevalence of the important vector, *Culex tritaeniorhynchus*.

Of 309 virus-isolations from mosquitoes, 307 were from *Culex tritaeniorhynchus*, the other two from *C. pipiens*, which indeed may have been only temporarily carrying the virus. Why *C. pipiens* was not more involved is obscure, as experimentally it is a very efficient vector.

The happenings could be divided into three periods. During the first the spring brood of mosquitoes hatched, coming from eggs laid by over-wintered adults, and multiplication of insects proceeded till maximum numbers were reached by the end of June. During this phase, viruses were not isolated. Then, from late June till September viruses began to be isolated from the insects. During July the nestling herons hatched out; these were very attractive to the mosquitoes and in them the Japanese virus readily multiplied. They thus acted as 'amplifiers' increasing enormously the amount of virus circulating between mosquitoes and the creatures they bit. Traps were baited with heron-nestlings, chicks and other species; by counting the mosquitoes it was possible to find out which were the most popular victims. It turned out that nestlings of black-crowned night-herons were the favourites. These in nature were the most exposed; their parents, being nocturnal as the name implies, left their infants uncovered during the flight-times of the mosquitoes. These did bite chicks in nests high up in trees but more infections occurred when nests were at lower levels.

The herons in these colonies outside Tokyo flew from time to time to places nearer the town and, carrying the infection with them, were responsible for establishing foci of infection near centres of dense population. Near one of the heronries were kept many pigs housed in styes unprotected from mosquito attack (plate 23). Studies made with baited traps showed that *Culex tritaeniorhynchus* preferred pigs even to black-crowned night-herons. Moreover, samples of blood from pigs yielded virus and it was shown by the development of antibodies that during August ninety-eight to one hundred per cent of pigs had been infected. Infections were silent ones, just as in the herons, though there is evidence that the virus may be the cause of abortions in sows. Apparently *C. pipiens* does not readily bite pigs; this may account in part for its failure to play a significant role, at least in Japan.

Man becomes infected with the Japanese virus by the bites of mosquitoes in the urban foci referred to above and also by those frequenting his piggeries, for he is in much closer contact with pigs than with herons. Pigs are thus 'link hosts'. We thus have reservoir or maintenance hosts in adult herons, amplifying hosts in the heron nestlings, link hosts in pigs and incidental hosts in horses and man, which are probably dead-ends as with the American equine viruses. Only in the last group do we find evidence of disease.

It is noteworthy, and was at first confusing, that the human disease, occurring mainly in August and early September, comes so much later than the peak of mosquito-numbers. It is now clear how this comes to pass. The mosquito-peak in June is followed by a peak in the number of *infected* mosquitoes after the heron nestlings have hatched in July. The 'amplification' of virus then causes a spillover to pigs and man in August, lasting until September.

Fortunately infection of man does not always lead to encephalitis; when this occurs it is similar to that caused by the equine viruses; permanent damage to the nervous system is not uncommon. Mostly, however, the virus infects children who have no symptoms or only those of mild fever. Five per cent of children between six and twelve are estimated to contract the infection every year and those who already have antibody may be reinfected. According to one estimate only 1 in 500 to 1 in 1,000 of these will suffer from encephalitis and even lower figures have been given. Even so, the disease is a serious one for Japan, for Tokyo is a city with such a huge population that, even though the percentage incidence of disease is low, the total number of cases is disturbingly high. In one epidemic there were eight thousand. How the Japanese encephalitis virus overwinters is just as much a puzzle as with the viruses discussed in the last chapter and the same theories have been put forward – and mostly discarded. At present the likeliest explanation seems to rest on the survival of some virus in overwintered *C. tritaeniorhynchus*. There may, however, be other vectors, concerned with adult rather than with nestling herons. Survival of virus in hibernating bats has been considered a possibility.

It is now being increasingly recognized that the disease is widespread over much of Asia from eastern India to eastern Siberia, though the ecology has not been as thoroughly studied as in Japan and it may not be the same everywhere. In far eastern Siberia the virus is associated with brackish pools on the coast and with other species of *Aëdes*. In Malaya the disease was brought to attention by the occurrence of serious outbreaks of encephalitis among race-horses. *C. tritaeniorhynchus* serves as a vector here, as also does *C. gelidus*, but a puzzling fact is that antibodies in birds are rather infrequent; possibly the right bird has not yet been tested. On the other hand pigs and not birds could furnish the virus-reservoir in a tropical country such as Malaya; there would be no need to hunt for an overwintering mechanism. In India another mosquito, *C. vishnui*, seems to be concerned; as in Malaya, the reservoir host was not identified. There was suggestive evidence pointing to a possible man-mosquito-man cycle, at least during part of the year. Clearly, every country or area has to establish for itself what is the local virus natural history: the Japanese virus seems able to adapt itself to different vectors, different hosts and different ecological conditions.

In 1917–18 an epidemic of encephalitis broke out in Australia; mortality was seventy per cent. Dr Cleland and his colleagues [42] isolated a virus, but this was lost before its nature was determined. Lesser outbreaks occurred in 1922, 1925 and 1926, then no more for twenty-five years. The mysterious malady was called Australian X-disease.

In 1951 encephalitis reappeared in the Murray and Darling river valleys in eastern Australia. This time the virus was definitely corralled and it proved to be a B-arbovirus, related to Japanese encephalitis: it was christened Murray Valley virus. There is a lot of circumstantial evidence for the view that this virus was the cause of Australian X-diseases. As with the other members of this group of viruses, birds are probably the normal maintenance hosts, man and other mammals being infected incidentally and, as a rule, inapparently. Several mosquitoes may act as carriers but *Culex annulirostris* is considered to be the main vector. The area involved seems in the ordinary course to be northern Australia including the northern territories and the north of Queensland, also New Guinea. In parts of the northern territories eighty to ninety per cent of aborigines were found to have antibodies. It has been suggested that presence of Murray Valley virus here, with consequent widespread immunity, may serve as an 'immunological barrier' preventing the spread southwards of Japanese encephalitis virus. Similar barriers may serve to prevent other viruses of this family from invading each other's territory.

The 1951 and other outbreaks of infection in south-eastern Australia have coincided with seasons of exceptionally heavy rainfall further to the north. It has been suggested that these conditions, by causing abnormal flooding, may have encouraged abnormal southward migration of water-birds carrying the infection, and also favoured unusually successful breeding of mosquitoes [117]. It must be admitted, however, that in the last few years cases have turned up in south-eastern Australia at times when the supposedly favourable weather conditions were not in evidence: evidently there is more to be learnt yet. Nevertheless outbreaks in Victoria and New South Wales very probably represent a spill-over of the virus from its endemic areas in the tropics; there of course mosquitoes would be active all the year and problems of overwintering would not arise. Study of the existence of virus-infection in little pied cormorants and other water-birds was made easier by a simple trick. In many mammals antibody passes through the placenta into the blood of the foetus, so that baby animals are born with some maternal antibody, which protects them from infection while they are very small and vulnerable. It only persists for a short while, but tides them over till their own defence mechanisms can get to work. In a similar way antibody reaches a chick from its mother by way of the egg-yolk. It was therefore possible for scientific bird-nesters to obtain samples of eggs from various species, to examine the yolks and so determine which birds and how many of them had been exposed to the virus [178].

We now cross the Pacific ocean to encounter the St Louis encephalitis virus, which is widespread in western and mid-western North America. Big epidemics occurred in the mid-west in 1932 and 1933 and the virus was isolated soon after. Outbreaks have occurred intermittently since then with small numbers of cases between whiles. A widespread outbreak occurred in 1964. Cases occur chiefly in late summer, doubtless for reasons similar to those made clear in studies of the Japanese virus. Once again it seems that we are dealing with a bird virus spilling over to affect man in years when it has flourished unusually well. Most affected people have symptomless infections or just a short fever: only the unlucky ones get encephalitis. The vector in the west is mainly *Culex tarsalis*, the mosquito which also carries the western equine encephaliis virus; in the mid-west *C. quinquefasciatus* and *C. pipiens* seem to take over. Yet other *Culex* species seem to be concerned in Trinidad. The distribution amongst wild birds is not thoroughly understood, but it is likely that chickens, house-sparrows and other birds frequenting the neighbourhood of houses are important link-hosts helping the carry of virus across to man. The problem of over-wintering of virus is as difficult as for the western equine virus, which we have already considered (see p. 82).

Chapter 10

Dengue:
a Changing Tropical Disease

IN DENGUE we encounter a widespread virus infection of the tropics; this has apparently been changing its character, first in its epidemiological behaviour, more recently in the severity of symptoms produced. It is due to arboviruses of the B group but differs from the cluster of viruses dealt with in the last chapter in that it affects only mammals, not birds and is transmitted by day-biting mosquitoes of the genus *Aëdes*. Though there is overlap in serological tests with those viruses, dengue does not lie as close to them as they do to each other. There have been two serological types of dengue recognized for some years; recently two more were discovered, and now yet another two may have to be added.

Dengue gives rise to a very sharp unpleasant fever with severe pains in head, limbs and back, sweating and rashes. It has not been considered a very fatal disease, but, as we shall see, a more serious form of the disease, haemorrhagic fever, has been recognized recently. Its headquarters are in southern and eastern Asia but it is known also from southern parts of North America, the Caribbean, the Mediterranean and Australia. Into some of these areas it may have been introduced in historical times.

The important vector in urban areas is the black-and-white yellow-fever mosquito, *Aëdes aegypti* (plate 20), and it is the relation between virus, mosquito and host which gives the clue to the behaviour of dengue in the world [151]. For the parts of the world where dengue occurs may be divided into two – endemic areas, where the virus is present all the time, and epidemic areas, into which the virus breaks in periodically. It has been shown that *A. aegypti* is much affected by temperature. It bites most freely between 79° and 95°F and ceases to bite at temperatures below 68°F – or according to another author 59°F. Exposure to 43°F for twenty-four hours kills the adult females, which are the ones of importance. Furthermore mosquitoes held at 60°F during the extrinsic incubation period (see p. 78) fail to become infective, for virus does not multiply in them. From all these things it appears that at temperatures round about 65°F, transmission of dengue will cease [47]. All this fits in with what is shown in the map (figure 14).

The solid lines show the 64°F isotherms for mean monthly temperatures in the winter – January in the northern, July in the southern hemisphere. Within these lines are most of the scattered outbreaks to be seen in an endemic area, where virus is present throughout the year. The dotted lines indicate the corresponding summer 64°F isotherms. Between the lines for summer and winter are the areas in which *A. aegypti* can flourish and

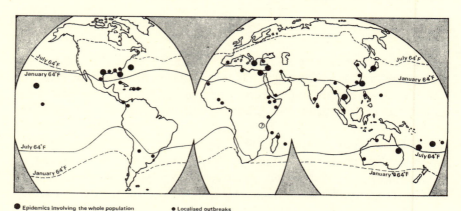

Figure 14. Map showing how localized outbreaks of dengue mostly occur inside the 64° winter isotherms and major epidemics outside those lines.

therefore transmit dengue only during warmer months. This will only happen intermittently and populations therefore lack the immunity seen in more tropical countries. Accordingly, in this zone, or near it, occur the infrequent big dengue outbreaks. The biggest of these have each involved some millions of people and have occurred in the southern United States (1922), Australia (1925–6), Greece (1927–8) and Japan (1942–5).

Aëdes aegypti is mainly a domestic mosquito, breeding in small collections of water in the neighbourhood of houses. A man-mosquito-man cycle is very easily established to explain the behaviour of dengue over much of the world and it is unnecessary to look any further. Its occurrence in tropical and subtropical countries, as the map shows, is almost world-wide. It is, however, not certain that what has been called dengue in parts of Africa may not be an infection with some other arbovirus. The two serological types, 1 and 2, have both been widespread, while 3 and 4 have come mainly from the Philippines and Thailand [71]. An extensive outbreak in the Caribbean area in 1964 seems to be caused by a virus related to type 3. There seems to be a fairly solid immunity against a second infection by the same serotype, but resistance is present even against a different type for a month or two.

Sir Macfarlane Burnet has suggested [34] that some of the difficulties in understanding the epidemiology of dengue could be resolved if there were antigenic changes going on, similar to those described for influenza as 'antigenic drift'. Firm evidence for this notion is, however, not forthcoming at present and the facts may be explained by the other phenomena discussed in this chapter.

Gordon Smith [151] has drawn attention to a change in the epidemiology of dengue which has occurred in Malaya since the beginning of the century. Before 1900 there were larger outbreaks involving whole populations; Malaya behaved like the epidemic areas just discussed. But since about 1900 the disease has behaved as though endemic in urban areas with only smaller scattered infections. Much the same may have been the case in other parts of south and south-east Asia. After discussing various possibilities Gordon Smith comes down in favour of the view that dengue used to be a rural disease transmitted by mosquitoes related to *Aëdes albopictus*, but that at the turn of the century *A. aegypti* was introduced, probably from Africa, thrived in the towns and showed itself to be a highly efficient vector of dengue. In the towns thereafter the endemic situation became established, with all too much tiresome illness due to dengue, but no longer any very big epidemics. Visitors from non-endemic areas have been, however, particularly likely to be attacked. There is evidence that *A. aegypti* is still extending its range in Asia and we may well see resulting changes in the epidemiology of dengue in other countries.

If this view of dengue is correct, what of the natural history of rural dengue? Is there a reservoir in wild animals or birds? Examination of sera from various species could be expected to provide an answer. Care would be necessary to avoid being led astray by cross-reactions with other B arboviruses, such as the widely prevalent Japanese encephalitis virus. It turned out that many domestic animals had antibodies against the Japanese virus but not against dengue; such as there were could be explained by these tiresome cross-reactions. Birds, also, and ground-living forest animals such as rats could be dismissed as unlikely to be important. On the other hand a number of tree-top dwellers had antibodies to dengue types 1 or 2 and not to Japanese virus. Though comparatively small numbers could be tested, seventy-three per cent of monkeys, mostly Macaca irus, fifty-eight per cent of slow lorises, twenty per cent of squirrels and seven per cent of civet cats were positive; so also were some flying foxes. Dengue may be, then, naturally a virus of the tree-tops, jungle dengue, comparable with the jungle yellow fever we shall meet in our next chapter. What is the natural vector in the tree-tops we do not know, though it is probably an *Aëdes* of some species. Maybe it is one of the group of closely related mosquitoes related to *A. albopictus*; members of this group are known to be dengue-vectors in Pacific islands and elsewhere, sometimes when *A. aegypti* is absent, or maybe it is another tree-top *Aëdes*, with *A. albopictus* and its friends acting as link-vectors transferring infection from the tree-tops to people on the ground below. We shall meet examples of this in discussing yellow fever.

Now to the story of dengue must be added a new and disturbing chapter. In 1954 there appeared in Thailand and almost simultaneously in the Philippines a disease which came to be called haemorrhagic fever. It has appeared subsequently in Malaya and the south of India. Its chief incidence was in children, in whom there was considerable mortality, especially in those under one and between the ages of three and six. Further it attacked only orientals; those of Caucasian origin were usually spared. There were seen small haemorrhages in the skin (purpura) and haemorrhages from the gut and elsewhere; mortality was due to shock with failure of circulation. In the Philippines Dr Hammon and his colleagues isolated viruses related to but distinct from the previously known two dengue serotypes; these were called dengue types 3 and 4 [71]. They were associated particularly with this haemorrhagic disease rather than with classical dengue. In Thailand a number of different viruses were isolated, dengue types 2 and 4 and other types related to 1 and 2 but perhaps qualifying to be regarded as new types, 5 and 6. What was very surprising was that other cases yielded a virus previously only known from Africa, Chikungunya; Asian strains of that differed a little from African ones. Outbreaks due to this virus were described on p. 85: in Africa the haemorrhagic manifestations were not evident. In still other Thailand cases no arbovirus was isolated. The outbreaks seemed to be urban in character and transmitted almost certainly by *Aëdes aegypti*, but whether or not there is an animal reservoir for the new types is quite unknown. The whole thing is still mysterious and several possibilities must be considered. Perhaps there have been mutations amongst the dengue viruses with antigenic changes going hand-in-hand with an increased tendency to damage blood-vessels. Isolations of several different viruses from these cases, or none at all, suggest the possibility that an unidentified virulent agent may be being transmitted to the sufferers along with the better-known viruses, and that this may be causing the haemorrhagic disease. Yet again it may be that something in the environment is causing several previously fairly harmless viruses to produce these fatal infections.

We do not know the answer, but must bear in mind some happenings in other parts of the world. Haemorrhagic fever was really brought to the attention of the western world in 1953 when it affected many of the United Nations troops in Korea [60]. Symptoms were like those found later in the Philippines and Thailand but with kidney failure much to the fore. The disease had been known to doctors in the East previous to this time. The infection has been transmitted to volunteers but, despite tremendous efforts, teams of American workers wholly failed to infect any laboratory animal or grow a virus in culture. Circumstantial evidence suggested strongly that the reservoir was in small rodents amongst which the virus was probably transmitted by mites [172]. Volunteers have been infected by injecting suspensions of mites. In contrast to other arbovirus infections this disease is commoner in winter, probably because field rodents gather for shelter in or near houses at that time. Infection of man may occur through contact with rodent faeces.

The disease is reported also from a number of areas in the USSR. Here we come up against another complication. Some of the haemorrhagic fever there seems to be due to this unidentified agent, but some of it is associated with infection by arboviruses of the B group transmitted by ticks. Of these we shall encounter Omsk haemorrhagic fever later (see p. 110) and another haemorrhagic tick-borne disease in India, Kyasanur forest disease (see p. 111) [184]. Nor is this all. Somewhat similar haemorrhagic diseases have turned up in Argentina [114] and more recently still in Bolivia [103]. The Argentinian one is due to a virus not in the A or B group and the Bolivian one is related to it. In both these instances there appear to be associations with rodents and mites.

It is possible that the sudden appearance of these diseases in different parts of the world is an illusion. It may well be that when attention has been drawn to haemorrhagic fevers in one area, people elsewhere begin to realize that they are seeing something similar. It is nevertheless very puzzling to find a number of different viruses, some of them quite unrelated, causing rather similar serious disease in some parts of the world, while behaving in the normal way in other areas.

Chapter 11

Yellow Jack

THE REMARKABLE STORY of the dreaded yellow Jack or yellow fever involves matters of much importance in history [23, 162]. The tale unfolds first in Africa, then in America, and we shall see how, as the natural history becomes clearer in one continent, similarities and differences are revealed in the other. As in the case of dengue, there are important differences between endemic and epidemic areas. In West and Central Africa yellow fever has doubtless existed from time immemorial. Where the disease has thus been endemic, the native Africans have had a high degree of resistance. Outbreaks of serious disease among them have been comparatively rare and most have sustained immunizing infections in childhood, often in the absence of any clinical evidence that the virus was around. When, however, Europeans first set foot in West Africa the results were very different. Yellow fever quickly took its toll. High fever, vomiting and severe pains were early symptoms, not so very different from dengue; but then, often after a period of remission, came jaundice, black vomit, delirium and frequently death, the liver being fatally damaged. The diagnosis was often confused with that of malignant malaria: the two diseases together soon gave West Africa the title of the white man's grave.

Nevertheless trade between Europe and West Africa was brisk and this led to the introduction of yellow fever particularly to Spain and Portugal. In 1857, six thousand people died in Lisbon in the course of a short epidemic. The infection even reached South Wales where a small outbreak led to thirteen deaths mainly among dock-workers. It was, however, in the western hemisphere that the most violent epidemics occurred. Not only Central America and the Caribbean area suffered, but ships carried the disease to the southern United States with epidemics occurring on a number of occasions as far north as New York, Baltimore and Boston. An epidemic in the Mississippi valley killed thirteen thousand people and brought business to a standstill.

Yellow fever affords an outstanding example of how knowledge of the natural history of an infection teaches us how to control it, even when the causative agent is still unknown. Carlos Finlay had a Scottish father and a

French mother; he was educated in Europe and the USA and practised medicine in Cuba. A team of investigators from the USA visiting Cuba had previously stated that the agent transmitting the disease must be in the air, a suggestion which turned out to be true, though not in the way they had thought. Carlos Finlay spotted that a mosquito was the 'agent in the air' and even suggested that *Aëdes aegypti* was the insect in question. For a long time he was disbelieved, but the truth became accepted some twenty years later as a result of work carried out by a commission under Dr Walter Reed. Some years later it was discovered that the causative agent was a virus, now known to be another member of the B group of arboviruses.

When the French engineer de Lesseps, who had built the Suez canal, attempted to construct the Panama canal also, he met with complete failure, a principal cause being the illness in his labour gangs caused by yellow fever and malaria. The canal was only built when Major Gorgas controlled the death-bringing mosquitoes as a first step in the operation. *Aëdes aegypti*, as we have already seen, is a domestic insect, breeding in water-containers around houses, and it is therefore not difficult to impose sanitary rules which will eliminate it.

It was believed for some time that all facets of yellow-fever epidemiology could be explained in terms of transmission between man and *Aëdes aegypti*, for no other vertebrate and no other mosquito seemed to be concerned. So the Rockefeller Foundation carried out a programme for eliminating yellow fever from the American continents by destroying the vector mosquitoes. Their workers evolved the concept of key centres. Any infectious disease will die out when the parasite has reduced the susceptibles to a sufficiently low level by either killing or immunizing them. This was referred to as 'the spontaneous elimination of yellow fever by failure of the human host'. It was argued that yellow fever could only persist in a centre of population which was either large enough to furnish continuous fresh supplies of susceptibles through the birth of children or where there was a regular arrival of immigrants. If such key centres could be freed from infection there should be no spread to smaller centres where the virus would be at most transiently prevalent and the disease should be eliminated. A minor difficulty arose in that there were some wandering rural outbreaks with virus being carried around from village to village, not reappearing in any one place until new susceptibles had grown up. Nevertheless it appeared for some years that the tremendous efforts of the Rockefeller Foundation had been crowned with success.

Then fell the blow and the disillusionment. Cases of yellow fever began to turn up in South America and the Caribbean in places where there were no *Aëdes aegypti* [159]. A clue was afforded when a lady in Trinidad told Dr (later Sir Andrew) Balfour that old negroes in her employ said they could foretell a yellow fever outbreak by the finding of dead and dying howler monkeys in the nearby woods [23]. It has indeed been gradually revealed that there exists what is called 'jungle yellow fever', an endemic infection in monkeys and possibly other species in the South and Central American jungles. Moreover this is transmitted, not by *A. aegypti* but by tree-top mosquitoes, particularly species of *Haemagogus*. The incidence in man is quite different from that of urban yellow fever, adult males being mainly affected and particularly foresters and others whose work brings them into close contact with the jungle.

Haemagogus mosquitoes normally live only in the tree-tops and breed in collections of water in tree-holes. Infections of man have occurred particularly when foresters have brought a big tree crashing down, thus setting loose the *Haemagogus* at ground level among the forest workers (see figure 15). These moquitoes may, however, occasionally become very abundant and take to breeding in domestic barrels and tanks and entering houses; this happened at the time of an epidemic of yellow fever in Bolivia.

Cases of jungle yellow fever are mainly sporadic, but the infected men of course return to their villages and there, if *Aëdes aegypti* is present, they may serve as the focus of origin for a local outbreak. The discovery of this jungle yellow fever has of course involved much disappointment, for elimination of infection from monkeys and mosquitoes in the tree-tops is something almost impossible to achieve.

Something has been learnt about the jungle reservoir of the disease. Almost all the South American monkeys are susceptible to infection, but especially spider monkeys, red howlers and capuchins; also marmosets. *Haemagogus* are day-biting mosquitoes, and it is thought that the monkeys may be bitten particularly when, as is their habit, they take a little nap at mid-day. Otherwise they are too alert to let themselves be bitten.

In contrast to what we shall find when we consider African monkeys, some of the South American species are far from being wholly resistant. As already mentioned, red howler monkeys may be picked up dead in the forests at times when yellow fever is active. As in the wandering rural outbreaks in man, so it appears that yellow fever epidemics, or enzoötics, amongst monkeys tend to move across country. The occurrence of human cases in successively infected areas is evidence of this; odd infections of man serve as indicators of more extensive disease among monkeys. There may, however, be more direct evidence; finding dead monkeys gave

Figure 15. Yellow-fever cycles in South America.

indications of how yellow fever was steadily moving northwards into Central America in 1948–55. It apparently began in Panama and successively involved Costa Rica, Nicaragua, Honduras and Guatemala, only moving during the wet season when *Haemagogus* were active. Yellow fever appeared in Trinidad in 1954, having, so far as anyone knew, been absent for forty years. Another spreading epidemic apparently started in the interior of Brazil in 1933 and spread south and east in the course of several years.

How do these wandering outbreaks manage to spread? Bands of monkeys move about in the forests but only over a few kilometres. One epidemic wave, however, is estimated to have travelled at a rate of about two hundred kilometres a month, much too fast to be explained by monkey-movements. Far more probable is it that infected mosquitoes are carried by the wind. Marked mosquitoes have been recaptured $11\frac{1}{2}$ kilometres from their point of release and it is reasonable to suppose that some may have been carried very much farther than that.

The problem of how yellow-fever virus overwinters is not as difficult as with arboviruses in temperate countries. Nevertheless, there is a problem. There are dry seasons lasting for a few months when mosquitoes are fewer and when yellow fever is not obviously active. Very possibly the virus persists in some surviving mosquitoes. It has once been recovered from a *Sabethes* mosquito which is known to survive as an imago during colder and dryer months. Some workers have suspected that just as the *Aëdes*-man cycle is superimposed on a *Haemagogus*-monkey cycle, so that in turn depends upon an underlying cycle in some other vertebrates and vectors. Indeed there is evidence that jungle yellow fever may occur in the absence of monkeys. Marsupials, especially four-eyed opossums (*Metacheirus*), may be infected and it is thought that they may well serve as maintenance hosts in some areas.

Several facts have served to convince most workers that yellow fever had its origin in Africa and was introduced into the New World through human agency, perhaps at the time of the slave trade. *Aëdes aegypti* is thought to be of African origin and similarly introduced. The liability of many New World monkeys to fatal yellow-fever infection contrasts with the natural resistance of African monkeys; this points strongly to a longer association between the virus and the monkeys in Africa. So, too, the African negroes show a high degree of resistance, and this is not shared by South American Indians.

In Africa the distinction between urban and jungle yellow fever is not quite as sharp as in South America. *Aëdes aegypti* in Africa is not a wholly urban insect, breeding also in tree-holes in the jungle, its probable original habitat; and accordingly it may be present almost anywhere. Yellow fever seems to be naturally a disease of the monkeys in the tree-tops. Evidence of infection from presence of antibodies in the sera is to be found in almost all the African species: the red-tailed and *Colobus* monkeys are referred to as being affected particularly often. The grivet monkey, *Cercopithecus aethiops*, is less frequently infected, probably because its habits are largely terrestrial. It is now well established that the mosquito *A. africanus* is the main vector, and this is a nocturnal tree-top mosquito which probably bites the monkeys when they are asleep at night.

The discovery of the importance of *Aëdes africanus* rests on a curious story. Sentinel monkeys, placed in cages in the tree-tops in likely areas where *A. africanus* abounded, just refused to become infected. It was finally discovered that this mosquito, though fully able to pass through the bars of a cage, never in fact did so. When the monkeys were tethered in the open the story was quite different: the mosquitoes bit and infected them with the virus. As in South America the yellow fever virus probably exhausts the supply of susceptibles in any one area and so the infection is probably always on the move around the jungles. There is less evidence than in the New World as to its speed of travel. It is, however, likely that wind-borne mosquitoes play a part since there has been evidence of spread for a distance of twenty miles to an island in Lake Victoria.

In Africa we again meet the problem of how virus persists during drier seasons. It may well be that some *Aëdes* survive through unfavourable times and that no further explanation need be sought. It has, however, been suggested that infection in monkeys may die out and have to be renewed from some more permanent source. Bats have been suggested as being possibly concerned: antibodies have been found in them and they could also serve in the wider distribution of infection. Attention has also been paid to bush-babies (*Galago*) but the evidence for incriminating them is not good.

So much for yellow fever as a jungle-monkey disease, but how does it get from the African jungle to man? The mechanism seems to be rather different from that in South America [158]. There is no evidence that *Aëdes africanus* descends very often from the heights to taste human blood. The monkeys, however, do come down from their trees in order to raid the adjacent plantations, especially in search of bananas. Here they may be bitten by another mosquito, *A. simpsoni*, which also bites man. *A. simpsoni* thus serves as a very efficient link-vector to carry the virus from the jungle to the village. In the village *A. aegypti* may be present also and so we have the makings of a local outbreak in man. So long as only Africans are

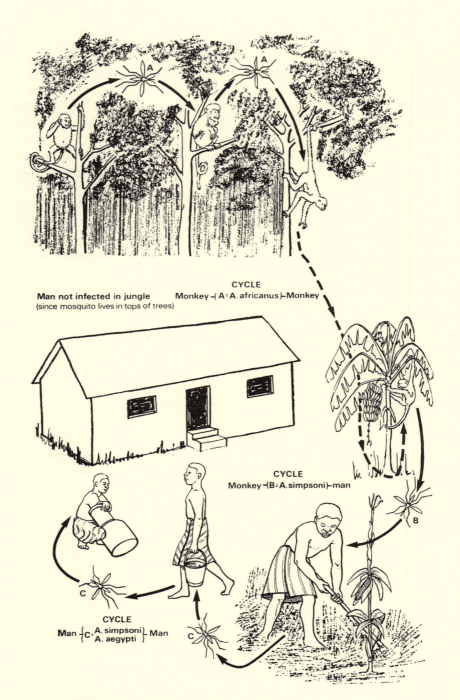

Man not infected in jungle
(since mosquito lives in tops of trees)

CYCLE
Monkey ⟶(A∶A. africanus)⟶Monkey

CYCLE
Monkey ⟶(B∶A. simpsoni)⟶man

CYCLE
Man {C∶ A. simpsoni / A. aegypti } Man

Figure 16. Yellow-fever cycles in Africa.

present, the 'outbreak' will often be a wholly silent one, only revealed when people are bled and discovered to have developed antibodies.

The distribution of yellow fever is often puzzling. There may be places where there is yellow fever carried by *Aëdes africanus* in the tree-tops and where *A. simpsoni* is present at ground level and yet there is no infection in the nearby villages. In one area it was observed that a wide band of elephant-grass separated plantations from the jungle and this kept raiding monkeys away from the bananas. In another part the *A. simpsoni* belonged to a race which took no interest in human blood. So for two different reasons infection failed to spread. Although in endemic areas Africans are highly resistant, the same is not true for Africans elsewehere. Two very serious outbreaks have occurred in recent years. In 1940 an enormous outbreak occurred in the Nuba mountains in the Sudan. There may have been as many as forty thousand cases and there were certainly at least fifteen hundred deaths. *A. aegypti* was present in some but not all of the villages involved and other *Aëdes* species, particularly *A. vittatus*, probably played a part. Then in 1960–2 an epidemic spread from the Sudan into Ethiopia, which had been apparently free for a number of years. It was apparently halted by a vaccination campaign (see p. 204) but not before there were something like fifteen thousand deaths. Mortality in one area was as high as eighty-five per cent [142].

There has naturally been great concern as to the possible consequences of the introduction of yellow fever into southern Asia with its teeming populations [51]. We may know how to prevent the disease by vaccination, but to prepare and administer hundreds of millions of doses of vaccine within a very short time could present almost insurmountable difficulties. Why has yellow fever not already reached Asia as, apparently, Chikungunya has done ? There has for centuries been much sea-borne traffic across the Indian Ocean, much of it in dhows carrying water supplies in which *Aëdes aegypti* could doubtless breed. And if man carried yellow fever from West Africa to America, why not the shorter distance to the continent of Asia ? Nowadays, air traffic could carry an infected person from Africa to Asia in a matter of hours. That, of course, is the reason why there are strict internationally agreed regulations about yellow-fever vaccination of travellers in certain areas and for the spraying of aeroplanes to prevent possibly infected mosquitoes from escaping. The vaccination is probably easier to enforce and more effective than the spraying.

It has been shown that *Aëdes aegypti* mosquitoes from the Orient do not differ from African ones in ability to transmit the virus. One is strongly tempted to come to the cheering conclusion that if yellow fever could establish itself in Asia it would already have done so. The distribution of the disease in Africa is highly relevant. Its headquarters are in West Africa; it exists in Uganda but was for many years not actually isolated there; there is serological evidence that it may reach the east coast but it only just does so. The officially declared endemic area, according to the World Health Organization, covers a good deal of the east coast. This official view is just as well for one wants to be on the safe side. But good evidence that the virus extends so far is rather doubtful. In view of known cross-reactions with other B arboviruses, the intepretation of positive findings in tests of sera must be cautious, unless they are very definite. There is some doubt nowadays as to the significance of positive reactions in sera from places far south, such as Bechuanaland.

It seems that there is some sort of biological barrier preventing the virus from spreading too readily eastwards. The story of the elephant-grass barrier to the monkeys and of the *Aëdes simpsoni* which did not bite man shows us how much difference an apparently trivial thing can make: the habits of *A. simpsoni* may prove in fact to be all-important. We saw in chapter 9 how the viruses related to West Nile – the Japanese encephalitis, Murray Valley and St Louis viruses – tended not to invade each other's territories, perhaps because of an immunological barrier. It has been suggested that in a similar way yellow fever and dengue may be mutually exclusive, even though they are not so close to each other antigenically as to show cross-protection in laboratory tests. They do seem to inhabit different parts of the world on the whole: though both have occurred in Trinidad, even if not simultaneously. There has been suggestive evidence of mutual exclusion in mosquitoes, but to use this as an explanation one would have to imagine an impossibly high rate of dengue-infection in Asiatic *A. aegypti*.

All we can do is to admit that we do not know why Asia is free, to thank Heaven that it is so, to keep our fingers crossed and take every precaution to see that she stays free. It is likely that there is an unexplained biological explanation of her freedom, but until that is fully understood, it would be folly to take chances.

Chapter 12

Ticks

AFTER MOSQUITOES, the most important vectors of arboviruses are ticks, and those concerned belong to several different genera [153]. To understand their role we must know a little about their life history and will take an *Ixodes* as an example. They mostly live in fairly circumscribed habitats, for they are particular about conditions, and the humidity, in particular, must be just so. Within the right sort of area the female tick lays her eggs, a thousand or so, for the chances that any one will reach maturity are small. The little larvae, unlike the adults, have only six legs. They climb up blades of grass and wait for a passing small rodent to which they may cling on (plates 23 and 24). They then make their way to a bare spot usually around the head, dig in their proboscis and suck blood, usually for a day or two. When satisfied, they drop off wherever they happen to be. They then take their time in digesting the blood and moulting, to change into an eight-legged nymph. Then, up the vegetation again to wait for another victim. This time they prefer something a little bigger than a small rodent, perhaps a rabbit or a bird. The cycle is repeated as before and the tick, now attained to adulthood, looks for still bigger game, perhaps cattle, deer, foxes – or man. All this sounds very chancy, and so it is. The habitat where the ticks are to flourish must have not only the right physical conditions but be inhabited by victims of different sizes to suit the differing tastes of tick larvae, nymphs and adults. The gorged ticks may well happen to drop off their hosts in unsuitable areas and that is then the end of them. If the vertebrate host spends most of its time away from the ticks' favourite habitat, that sad event is particularly likely. When the tick population is a stable one, we may imagine that on the average only two individuals survive to carry on the next generation; but, with a thousand eggs laid, the mother tick can regard with equanimity the fact that 998 of her offspring fail to survive the many hazards which face them. People wanting to collect ticks for population studies or in search of viruses, commonly drag cotton-calico or other materials over likely vegetation; the ticks hang on, mistakenly supposing they have found a favourable host. Animal-skins wrapped round a hot water-bottle have even been used to make the deception more effective.

The largest group of tick-borne viruses is still contained in the big B-group of arboviruses. The viruses concerned are so close together according to serological tests that it was for long uncertain how many different ones there really were. Refined methods seem at last to have sorted them into viruses with discrete ranges and mostly carried by different tick-vectors [41]. We have louping-ill in Scotland, mostly known as a disease of sheep, carried by the tick *Ixodes ricinus*, the common sheep-tick; Central European tick-borne encephalitis, also with *I. ricinus* as its host; Far East tick-borne encephalitis carried by *I. persulcatus*; Omsk haemorrhagic fever; Kyasanur forest disease from India [184]; Langat virus from Malaya [152] and Powassan virus from North America. The viruses from Russia have been generally known as Russian spring-summer encephalitis but the name has been used for the Far Eastern as well as for the Central European virus or for the two together; so it is best to avoid the term. All these viruses are now said to belong to the tick-borne encephalitis complex. We will have a little to say about each of the viruses in the complex and then something about a few tick-borne viruses which are unrelated to the B arboviruses.

Louping-ill has the distinction of being the only arbovirus known to occur naturally in Britain. Few countries in the world are as free from arboviruses as we are. It occurs especially in hill-pastures around the Scottish border; also in Northern Ireland and to a lesser extent in South-west England. It affects sheep, less often cattle and pigs, and a few infections have been recorded in man, particularly laboratory workers. The disease in sheep is mainly inapparent, but may show up as a disease with incoordination going on to paralysis; losses amongst lambs may be serious. Whether the infection attacks or spares the nervous system depends often on the presence or absence of a very common and otherwise pretty harmless disease called tick-borne fever: this is due to a rickettsia. This organism apparently prepares the way for entry of the virus into the brain.

The vector, *Ixodes ricinus*, itself acts as a reservoir of infection, but the virus also circulates, without producing symptoms, among small rodents, grouse and deer. It has been possible to do a great deal towards controlling the disease in sheep by using a formalin-killed virus vaccine and by reducing the numbers of ticks by dipping. Encouraging results followed this course of action until 1960, when there were again disturbingly big outbreaks with mortality among lambs as high as fifty per cent. What apparently happened was this: vaccinated ewes had antibody which was passed to the lambs through the placenta (see p. 164). The lambs thus had a temporary immunity; they were soon bitten by the numerous ticks around them, but underwent symptomless infections with the temporary passive immunity from their mothers replaced in consequence by an active permanent immunity. But then the dipping programme brought the tick population right down, so lambs born later never received their naturally acquired virus dose at the time when their maternal antibody could hold it in check. Rodent populations are liable to tremendous waxings and wanings and with the waxings the ticks multiply, too. Probably as a result of such a change infected ticks become suddenly plentiful, with disastrous results to the lambs. Here is a good example of how knowledge of all the factors in the natural history of a virus must be understood, if interference by man is to have the results which are intended.

The tick-borne viruses of the USSR and northern and central Europe have been much confused. It now seems that there are two viruses, distinguishable on an antigenic basis, carried by *Ixodes ricinus* and *I. persulcatus* respectively. There may, as with so many other arboviruses, be inapparent, mild and severe infections, the Far Eastern form being likely to be much more severe with mortality in man as high as thirty per cent and with frequent occurrence of permanent paralysis. The Central European form is commonly biphasic, a preliminary fever being followed by a period of quiescence of some days; symptoms involving the central nervous system then begin.

Ixodes ricinus, inhabiting wooded areas, acts as a reservoir as well as a vector, for ticks once infected remain so for life. Furthermore infection may be carried over from one stage of the tick's life-cycle to the next, or even through the egg to the next generation – transovarial transmission. It has been claimed that at times infection in a female tick cannot be directly demonstrated, yet virus can be found in the young larvae hatched from the eggs which she lays. Transovarial transmission seems, all the same, far from satisfactory by itself as a means of carrying on the virus. It seems to be only ten per cent efficient, and if we consider how few tick-larvae ever reach maturity, it could obviously not suffice by itself. The normal hosts of the tick seem likely to be small rodents, birds and, for adults, larger mammals. Ordinarily ground-loving birds are most likely to be infected or fledglings of other species, just out of the nest. Birds are of especial importance, for migrants would be capable of spreading infected ticks for long distances; during the autumn migration, in particular, many birds may be carrying either ticks or virus or both. It appears that in ducks virus may persist for considerable periods. Lizards get bitten by ticks and can be infected with the virus, but only with heavy doses and are unlikely to be important. Bats may also have a role to play and there is experimental evidence that virus may be dormant in them while they are in hibernation.

They have, however, their own menageries of parasites and should not be very frequently exposed to terrestrial ticks such as *Ixodes*. Tick-borne viruses can be carried by mosquitoes, but they cannot apparently multiply in them – though some mosquito-borne viruses can grow in ticks. For mosquitoes to be effective in transmitting infection they must carry virus over mechanically from one victim to another, their full engorging meal having been interrupted.

A quite unexpected method of transmission was brought to light when virus was found in the milk of infected goats [176]. This had led to infection of numerous people drinking this goats' milk, the disease having been described as 'diphasic milk fever'. There may have been as many people infected thus as by being bitten by ticks. All the same, this is probably only another example of a blind-alley infection, the alley being just a little longer than in the case of the equine encephalitis viruses. The facts are important, however, as illustrating possible ways in which arbovirus infections could evolve into something for which arthropods were no longer essential.

Ixodes persulcatus is the tick of the taiga zone, the marshy northern forest, in parts composed of deciduous, elsewhere of coniferous trees, extending in a belt all the way from Scandinavia to the far east of Siberia. This is the zone where the Far Eastern tick-borne encephalitis virus holds sway. Normally it circulates among the small rodents, birds and other species of the taiga and no one is any the worse. But with the development of Siberia large areas of taiga have been cleared and the forest workers consequently exposed to the infection. Incidence is almost confined to them and does not readily spread to people with other occupations. We have seen that the requirements of many ticks are exacting, and so it happens that within the taiga there are only quite limited foci carrying serious danger of virus-infection.

The virus causing Omsk haemorrhagic fever is carried by ticks of a different genus – *Dermacentor*; in them, too, transovarial transmission has been demonstrated. Not much is known of its natural history, but it is of interest that musk-rats, a rodent established in but not native to the area, are liable to suffer from fatal epizootics. Moreover a number of workers handling infected musk-rats in the laboratory have become infected with the disease.

One of the most interesting tick-borne viruses of this group is the cause of Kyasanur forest disease; this turned up in 1957 as an apparently wholly new disease [184]. The main area involved was in Mysore province in southern India. Here, a number of dead monkeys were found in the forest, mainly black-faced langurs (*Presbytis entellus*) but also red-faced bonnet macaques (*Macaca radiata*). Cases of a serious haemorrhagic disease appeared in nearby villages; the villagers called it monkey disease and thought it was due to seeing or smelling a dead monkey. Stories of dead monkeys at once suggested that a much dreaded event had happened and that yellow fever had reached India. Fortunately this turned out not to be the case and the infection was identified as being one related to the tick-borne encephalitis group. Some of the monkeys were carrying ticks of yet another genus, *Haemaphysalis*, and it was soon established that the most important vector was *H. spinigera* (plates 23 and 24). It seemed probable that the dead monkeys and the human infections were simply indicators of a widespread infection in some other forest animal. Antibodies were found in some rats, squirrels, shrews and jungle-fowl. Inapparent infection of small rodents seemed a likely answer, as in the case of other viruses in the group. However, the rodents were mainly infected by *Ixodes* ticks, *Haemaphysalis* being relatively uncommon. In jungle-fowl, on the other hand, the *H. spinigera* were frequent. One unfortunate peacock carried as many as six hundred. A fair number of them also occurred on the three-striped palm squirrel, *Funambulus*. Transovarial transmission of virus was shown to be possible in *Haemaphysalis*. Cattle seem to be resistant to the infection. They may act as 'amplifiers', however, in another sense. *Haemaphysalis* ticks flourish on them and in their presence populations of this tick increase. They thus favour the spread of the disease in an indirect manner.

The great question is of course: where did this 'new disease' come from? Introduction by means of infected birds has been suggested. It is true that the disease has not been recognized elsewhere and that viruses carried by *Haemaphysalis* are not known elsewhere. On the other hand, there must be areas in the world sparsely populated and with little contact between the people and the jungle, where such an infection could exist and be long undetected. Alternatively the virus might have been present in the Kyasanur forest all along, the events being precipitated by involvement of a new tick with a habit of biting man and monkeys. Possibly also, adaptation to a new tick could be associated with a change in the virulence of the virus. There is some evidence that Kyasanur forest disease is extending its range southwestwards.

A further virus to be discovered in the group is the Langat virus from Malaya [152]. This is an infection particularly of forest rodents and is carried by *Ixodes granulatus*. This tick rarely bites man and the virus is of such low pathogenicity for man that investigations have been going on to see whether it might not be used as an avirulent live agent for vaccinating against its more dangerous relations; there is still doubt, however, as to whether it is quite harmless enough for such a purpose. Foci in the Malayan jungle where infected ticks can be found seem to be very circumscribed.

It seems curious that the taiga forests of Canada and Alaska have not provided tick-borne viruses as Siberia has done. There have been, however, a few isolations of a virus related to the others, the Powassan virus from Ontario and what seems to be the same agent from *Dermacentor* ticks in Colorado.

Colorado, too, is the centre of activity of the best-studied of the tick-borne viruses which are not related to the B-group of arboviruses. It is an arbovirus, but one standing in a rather isolated position. It is carried by the wood-tick *Dermacentor andersoni* and is fairly widespread in the moister parts of the north-western USA; it has also been recovered from Alberta, across the Canadian border. Infection of man leads to a severe fever, often of the saddle-back or biphasic type; involvement of the central nervous system is fortunately rare. Inapparent infections occur. Virus can be transmitted in ticks from larva to nymph and adult but occurrence of transovarial transmission is doutful; circulation of virus between larvae and nymphs of ticks and small rodents seems adequate to maintain the cycle. The rodents concerned are golden-mantled ground-squirrels and chipmunks, in both of which virus may be plentifully present in the bloodstream. Other rodents may be affected including Columbian ground-squirrels, but these animals are not adequate in themselves to maintain the virus-cycle; they are only found infected in areas where the golden-mantled ground-squirrels are present also.

There are other tick-borne viruses present in Florida, Japan, Bulgaria, Siberia, Uzbekistan, the Crimea and elsewhere in the USSR. The virus of Uzbekistan haemorrhagic fever is transmissible from man to man, apparently by the respiratory route. There has been very little systematic search for tick-borne viruses in the tropics – there are quite enough mosquito-borne ones to keep investigators occupied. However, Nairobi sheep disease, transmitted by a *Rhipicephalus* tick is well known as a cause of serious haemorrhagic gastro-enteritis amongst sheep and goats in Kenya, Uganda and possibly elsewhere in Africa. A rodent, *Arvicanthis*, may be the maintenance host.

Chapter 13

Sand-flies and Midges

BESIDES MOSQUITOES, some other families of biting flies transmit viruses amongst vertebrates. Not all families are concerned, however, even though they may be vectors of other infections such as worms or protozoa. The horse-flies, Tabanidae, seem not to carry any viruses, though *Chrysops* may transmit filarial worms; nor do the Muscidae, despite the habits of the dreaded tsetse fly (*Glossina*) in conveying the trypanosomes of sleeping sickness and numerous infections of cattle and horses. At the worst they may transmit viruses mechanically, particularly if their feeding is interrupted. No cycle of development occurs in these flies. The species of *Simulium* of the Simuliidae, sometimes called black-gnats or buffalo-gnats, may be a tremendous nuisance in temperate zones, and *S. damnosum* transmits a worm in the tropics; yet these too are, so far as is known, guiltless of carrying viruses. The two groups which we must still deal with are the genus *Phlebotomus* of the Psychodidae and *Culicoides* of the Ceratopogonidae.

Phlebotomus papatasi is the vector of sand-fly fever of the Mediterranean and Middle East. It is generally called sand-fly, yet elsewhere the term sand-fly is used for *Culicoides* midges; so it is perhaps safer to stick to the name *Phlebotomus* (vein-cutter or blood-letter) for the insect and to call the disease phlebotomus fever. The word papatasi, commemorated in the insects' scientific name, refers to the common designation, pappataci, of the fly in Italy and the Balkans.

The disease is pretty well coextensive with the distribution of the vector, that is from Italy, around the eastern part of the Mediterranean, the Black Sea, other parts of the central Asian USSR, the Middle East as far as western and central India. Numerous other species of *Phlebotomus* are not known to be concerned. *Phlebotomus* are hairy, sandy-coloured flies about 2–3 mm. long: the wings are raised and angled to form a V. They mostly bite indoors and at night. They settle on walls of houses and hop about, with pauses between hops, stealthily entering houses and eventually taking their blood meal. Because of these habits they are rather readily controlled by the residues of DDT spraying which get into the insects through their feet. Unlike that of most biting insects, their life-history does not depend upon water; their larvae live in nooks and crannies in masonry and earth, wherever a little organic matter has collected.

Seven to ten days after an infecting blood meal, possibly sooner, virus has multiplied in the flies sufficiently to allow transfer of infection to a second bitten person. Three or four days later he begins to suffer from his 'sand-fly fever' with pains in back, head and limbs, vomiting and other gastro-intestinal upsets, chilliness and sweating. Fever usually only lasts for three or four days but may recur during convalescence. In people native to the areas involved, infection is probably commonest among children and frequently inapparent; the more acute form of the disease is especially seen in people coming from elsewhere, for example, troops from Britain and the USA, immigrants and tourists. Second attacks are frequent and this is largely explained by the existence of two distinct serological types of virus. The several types of dengue were at least partly related and showed a little cross-immunity, but the two phlebotomus-carried viruses, known as the Sicilian and Naples viruses, show no cross-immunity whatever.

These things only became known after it had proved possible to infect suckling mice with these viruses by intracerebral inoculation, to develop a haemagglutination test and thus render the viruses available for laboratory study. Virus which has been propagated serially in mice becomes increasingly virulent for them, and simultaneously attenuated for man: there thus arises the possibility of a live attenuated virus for use in vaccinating man; it has not been widely used as the infection is easier to control by attacking the vector with DDT.

Phlebotomus flies are not in evidence during some winter months, and how the virus survives is unknown. There is no evidence as to possible existence of an animal reservoir. Two groups of workers have claimed that virus can be carried transovarially from one generation of flies to another: a third group has failed to confirm this. Possibly it is an infrequent event; only an occasional carry-over might suffice to keep the virus going. A suggestion has been made that the true reservoir is in mites which are found on four per cent of adult *Phlebotomus*, but there is no direct evidence in favour of this.

Fresh light on the whole question of the place of the phlebotomus-carried viruses in the world has come lately from Brazil [38]. Among the many viruses isolated from the Rockefeller Foundation's laboratory in Belém have been five which seem to have affinity with the sand-fly group. One, called Icoaraci, was isolated from a wild rat and was serologically related to the Naples virus. The others were serologically distinct, but are suspected to belong to the group as their behaviour in the haemagglutination test resembles that of the Sicilian and Naples viruses, but is rather different from that of arboviruses in general; they clump red cells best at 37°C and the haemagglutination differs in other ways also. The Brazilian viruses were obtained respectively from blood of a febrile man, from the organs of a sloth, from sentinel mice (see p. 77) and from a spiny rat. No direct evidence of association with *Phlebotomus* was reported by the Brazilian workers. We may hope that future research will make clear what is the significance of these puzzling findings and the relationship of these viruses to the *Phlebotomus* viruses of the old world.

The *Phlebotomus* viruses, so far as is known, resemble the generality of arboviruses in their properties, differing mainly in regard to vector. The next two viruses to be considered, those of blue-tongue, a disease of sheep, and African horse sickness, are probably not closely related to the others. They seem to be rather small spherical viruses with capsomeres, perhaps 92 in number, arranged round a core. They are relatively resistant to ether and may turn out to have affinities with reoviruses (p. 16). They differ also from arboviruses in general in being very labile as regards their antigenic make-up.

Blue-tongue has been known in South African sheep for many years. It may show itself as a brief uncomplicated fever or as a rapidly fatal disease with ulcerations involving mouth, tongue, nose and hooves. The cyanosis of the mucous membranes gives the disease its name. Sheep are mainly affected, but cattle may be attacked also; the disease is, however, milder in them and often symptomless, and never involves such large numbers. In 1943–4 blue-tongue appeared in Cyprus, a very virulent strain being concerned. Soon after it was recognized in Israel, Turkey, Spain, Portugal and Pakistan. In 1948 cases began to turn up in Texas, and in 1952 blue-tongue spread widely in California and in several other western states. Its wide dissemination in North America has suggested that it may have been present, unsuspected, for some time and not recently introduced.

In South Africa the disease has appeared after introduction of sheep into areas where neither cattle nor sheep had been previously present. A reservoir in some wild species may be presumed. Blesbuck, antelopes of the genus *Damaliscus*, have been shown to develop viraemia after inoculation. Sheep kept under cover at night have seemed to be safe from infection, so a nocturnal vector was thought to be concerned. It now appears that midges of the genus *Culicoides* carry the infection. These are the very small insects, sometimes known in America as sand-flies or 'no-see-ums' (plate 19). They are a nuisance in many countries, giving a small painful prick, which is followed by an almost immediate pink disc around the puncture-site and, in some people, a more prolonged itching spot or blister. In Africa several species of *Culicoides* may be concerned; in America *C. variipennis* has been particularly incriminated. With such minute insects proof that virus actually multiplies in the vector has been hard to obtain. It has been suggested that transmission may be merely mechanical, especially as blue-tongue becomes prevalent some time later in the season than the peak of *Culicoides*. However, we saw in chapter 9 how Japanese encephalitis came along some time after the mosquito-peak, waiting to be amplified by nestling herons. Very possibly blue-tongue virus also needs to be increased by some unknown amplifier. True biological involvement of *Culicoides* is rendered more likely since it has been shown that an extrinsic incubation period (see p. 78) of ten days is necessary before infected *Culicoides* will transmit the virus. The virus may persist in the body of a recovered sheep for four months, possibly longer, and in cattle, viraemia may be prolonged and intermittent. Carrying over of virus from one season to the next may be explained in such a way.

The blue-tongue virus exists in a number of serological races, a fact which has greatly complicated attempts at immunizing with vaccines. There may be as many as nineteen serological types. These do not turn up successively as with influenza; it seems rather that there are a number of antigens present in variable proportions in the strains present in the same general area. The different strains show variable amounts of cross-immunity, and it may be in fact hard to decide how many separate ones there are. Some success has been achieved with a vaccine made up of four strains which between them gave fairly wide coverage.

The story of African horse-sickness shows remarkable similarities to that of blue-tongue [128]. It, too, was first recognized in South Africa, but has since spread, it is carried by *Culicoides*, it is unlike arboviruses in general, both morphologically and in being relatively ether-resistant, and it is antigenically labile. The virus causes serious illness among horses, mules, and in the Middle East, donkeys. Acute cases die with oedema of the lungs; in more chronic ones the heart is involved and there is gross oedematous swelling especially of the head and neck; there may be overlooked, mild cases. The virus resembles that of blue-tongue morphologically, and the two must certainly belong to the same family. Evidence of transmission by *Culicoides* is of the same nature as for blue-tongue. In Turkey horse-flies (Tabanidae) and stable-flies (*Stomoxys*, of the Muscidae) have been accused of carrying the virus, but if so transmission is probably only mechanical. There is probably a reservoir in some African mammal; one naturally thinks of zebras, but these are said to be only slightly susceptible. The

reservoir might, however, be in a carnivore. Ferrets and dogs are susceptible: packs of hounds have suffered fatal infections after eating meat containing the virus.

In 1944 African horse-sickness spread to the Middle East, Turkey, India
and Pakistan, causing serious losses. It has, however, not, like blue-tongue,
turned up as yet in America. The variety of antigenic types repeats the
blue-tongue story; differences between them are not sharp and their numbers have been variously estimated as from eight to twenty-eight. A vaccine
made from eight selected strains is said to have given good results.

The last virus to be considered in this chapter, that of vesicular stomatitis,
is in several ways a sort of odd man out. It was first recognized in North
America in 1916 when horses were being brought from farms in the mid-
west for use with the armies in Europe. It caused in horses, also in cattle,
an infection which was readily confused with foot-and-mouth disease:
vesicles appeared on tongue and lips, though not commonly on the feet.
When it first became familiar, infection was most often seen in horses;
subsequently cattle have been more frequently affected. Infections were
only seen during the warmer months and died out with the coming of frost,
so an insect-vector was suspected. The symptoms are unlike those produced by arboviruses; further it seems to be a most versatile general-
purpose virus, capable of infecting not only most ungulates and rodents but
also ferrets and birds: inoculated geese develop secondary lesions on their
feet. It readily infects laboratory workers, causing either inapparent infections or an influenza-like disease. Fifty-four such infections occurred in one
laboratory.

It is an ether-sensitive RNA virus like arboviruses but morphologically
different. Electron-microscopy reveals bullet-shaped particles rounded
at one end and squared off at the other with a central hollow. One can also
see what look like caps, all ready to be fitted on to the virus's square end
[24] (plate 25). It is too early to know how to interpret these appearances, but it seems fairly safe to say that the virus is a representative of a
small family of its own.

There are two serological types known as New Jersey and Indiana; the
former is much the commoner and more virulent. It seems to be essentially
a New World virus: introductions into Europe and Africa have been quickly
stamped out. A number of different flies have been suspected as vectors,
horse-flies (*Tabanids*) and stable-flies (*Stomoxys*) in particular; their role is
probably only a mechanical one. Virus has also been recovered in Panama
from *Phlebotomus* caught in the wild. Furthermore the virus has been
shown to be capable of multiplying in *Aëdes aegypti*. It is not yet known

whether there is true biological transmission with multiplication both in vertebrate and in specific vectors, possibly mosquitoes; whether the virus is so catholic in its states that it can multiply in all sorts of insects as well as in all sorts of vertebrates; or whether much of the insect-transmission is merely mechanical. Transmission may also occur in the absence of arthropods, for instance amongst inoculated mice even in the absence of cannibalism. Deer in the wild apparently liberate much virus from ruptured vesicles and are suspected to act as amplifiers.

Recently in Trinidad a virus, Cocal, has been discovered, having antigenic relations to the Indiana strain of vesicular stomatitis. It was obtained from a pool of mites, *Gigantolaelaps*, removed from some terrestrial rice-rats of the genus *Oryzomys*, trapped in a swamp. Subsequent study [91] showed that antibodies to the virus were frequently present in the sera of trapped rodents; at one time fifty per cent were positive. Antibodies were also found in a few 'equine animals' but none in pigs or man. The *Oryzomys* did not show any illness, but rats of another genus, *Zygodontomys*, frequently died with paralysis after inoculation. They developed superficial crusts at the site of inoculation and it is possible that mites feeding on these crusts might transmit infection. The Trinidad workers suspected, however, that rodents might not be vital for the train of infection of Cocal virus. Its natural history has therefore still to be cleared up, as also has the question of its possible relation, in the course of evolution, to the virus of vesicular stomatitis. Recently it has turned up in association with a vesicular stomatitis-like disease in South America.

Chapter 14

Alternative Routes

OUR DISCUSSIONS of virus-transmission so far have perhaps over-simplified the matter. Respiratory viruses, such as colds and influenza, are easily thought of as those emitted from the respiratory passages and passing through the air to reach, usually rather directly, the noses and throats of other victims. So, too, the arthropod-borne viruses have to get from the blood-stream of one host, via the arthropod, to that of another. With the enteroviruses things were a little more complicated. Though polio virus and its relations multiply mainly in the intestine and are mainly passed out in the faeces, yet the pharynx is involved also, and transmission may involve both respiratory and intestinal routes. The impression must not be given that respiratory, enteric and blood-transmission covers everything. Some viruses can be transmitted mechanically or can make use of alternative routes, or again a combination of routes may be concerned. It is likely that methods of transmission are subject to evolutionary change; possible examples are mentioned in discussing tick-borne encephalitis (p. 110) and Q-fever (p. 200).

The poxvirus group affords examples of versatile viruses which can be transmitted in several ways. Infection of milk-maids' hands by cow-pox is doubtless mechanical, virus probably entering by minute points of damage to the skin. So, too, the milk-maids' hands certainly lead in a similar way to infection of other members of the herd of cows. The vaccinia virus used for human vaccination may be carried mechanically to infect other parts of the subject's skin or even to other persons, though this should not happen if proper hygiene is carried out. Molluscum contagiosum is a rather uncommon skin disease due to a member of the poxvirus family. Slowly developing nodules, becoming pearly white, are scattered over the skin; but it is only the skin which is involved and infection is almost certainly wholly mechanical by contamination with the contents of burst nodules.

Infection by smallpox virus itself is probably by the respiratory route. Early in the disease when the rash is appearing the saliva and respiratory secretions contain much virus and infection is a straightforward respiratory affair. At this stage virus in the skin-spots is lying deep. Here the virus is not a source of danger until the vesicles have developed and formed crusts. Virus from such crusts can dry up and survive for a long time. Smallpox patients are infectious as long as any crusts remain. Infection may occur through handling infected bed-clothes without any contact with the small-pox patient himself. Outbreaks in Lancashire have been traced to imported cotton. Smallpox is so greatly feared and is so infectious that smallpox hospitals have commonly been sited well away from built-up areas or even in ships anchored away from the land. There is some justification for this for the virus is believed to survive in the air and to be able to infect at a considerable distance. A case of smallpox infection is recorded where the only known source of infection was when the victim passed by a smallpox hospital on a bus. During a recent outbreak in England a number of cases occurred in circumstances where the source of infection just could not be traced.

A virus closely related to smallpox causes mousepox or infectious ectro-melia in mice. This word 'ectromelia' refers to the shedding of whole limbs – or the tail – as a result of gangrene following damage to local blood-vessels by the virus. This virus exists as a wholly latent infection in certain stocks of mice. It is latent, that is, until its equilibrium with its host is upset, usually by some manipulation or change in conditions brought about by a laboratory worker. It may then be lit up and cause very troublesome out-breaks of fatal disease amongst mice. The symptoms are very variable so that presence of the infection has often not been recognized and much con-fusion has resulted. The evolution of the disease in some outbreaks resem-bles that of smallpox in man, virus being shed from the dried-up crusts of rash which appears on the body. This rash is, however, not always seen and infection may then come rather from the respiratory tract. When an out-break has subsided, surviving mice may excrete virus for a long time in the faeces and virus may be recoverable from dried faeces or tails [62]. A role of ectoparasites has also been suspected but never really proved. One can hardly dispute that this is a versatile virus.

There is evidence that the pig-louse (*Haematopinus suis*) is of major importance in transmitting swine-pox in North America [146], but it seems that the disease can spread even when pig-lice are absent. It appears also that while fowl-pox may be transmitted by direct contact, mosquitoes of the genera *Aëdes* and *Culex* are very important in conveying infection. There is no evidence that the virus multiplies in the mosquitoes, but in one instance a mosquito was still carrying virus 210 days after biting an infected bird. Similar long persistence of virus on the proboscis of a mosquito is recorded also in the case of myxomatosis, another poxvirus, which will be

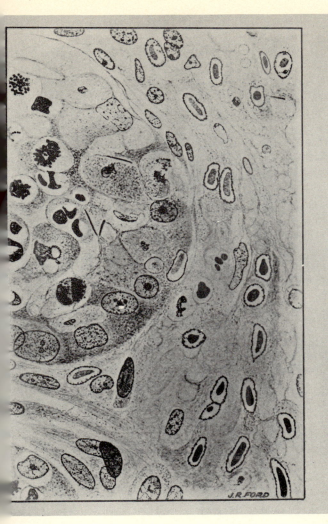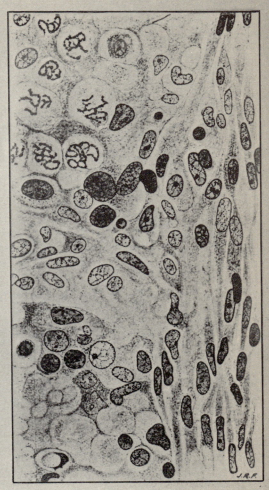

1. Sections of tissue-cultures of rabbit testis. On the right, normal, and on the left, infected with virus from rabbits; many cells show inclusion bodies in their nuclei similar to those found on chicken-pox and herpes.

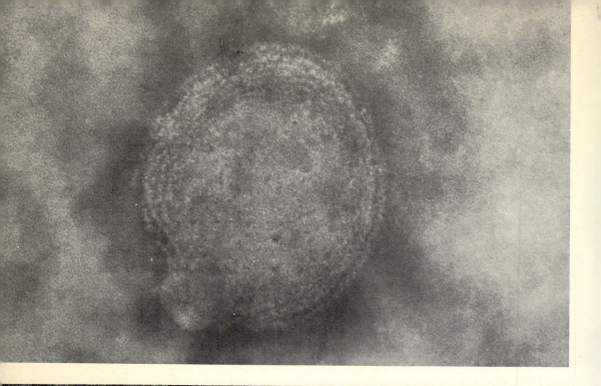

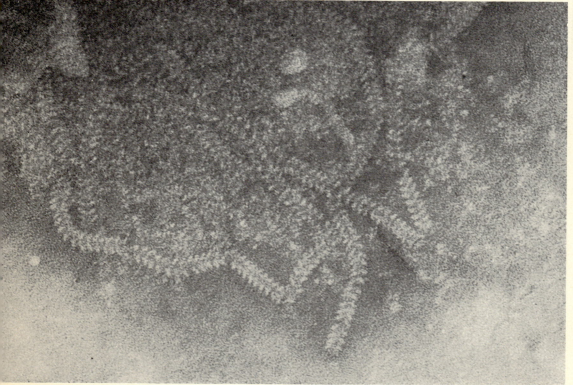

2. Electron-micrograph of a virus particle of Newcastle disease of fowls. The coiled internal structure can be faintly seen.
3. Helices released from the interior of measles virus (electron-micrograph).

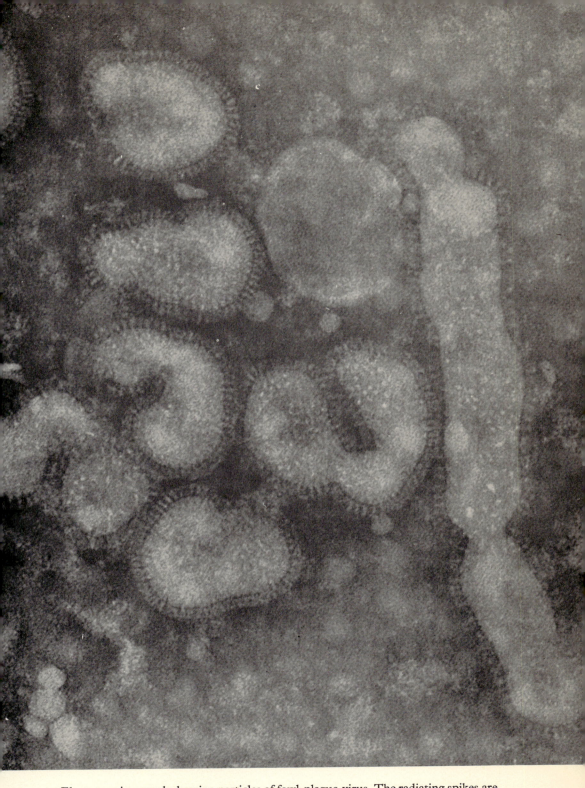

4. Electron-micrograph showing particles of fowl-plague virus. The radiating spikes are clearly shown on the surface of rounded particles and around one filamentous one.

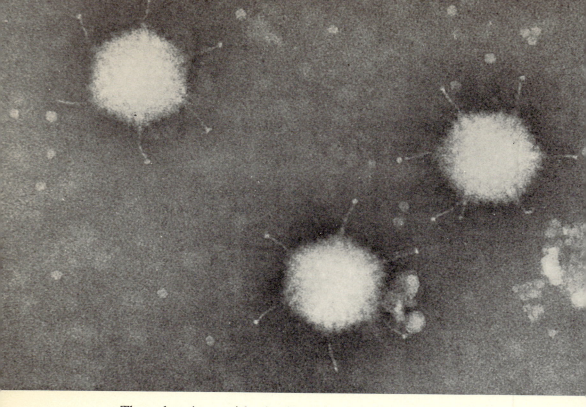

5. Three adenovirus particles showing surface structure and projections at each corner.
6. Numerous round virus particles from a human wart. The numbers of surface units are hard to count accurately.

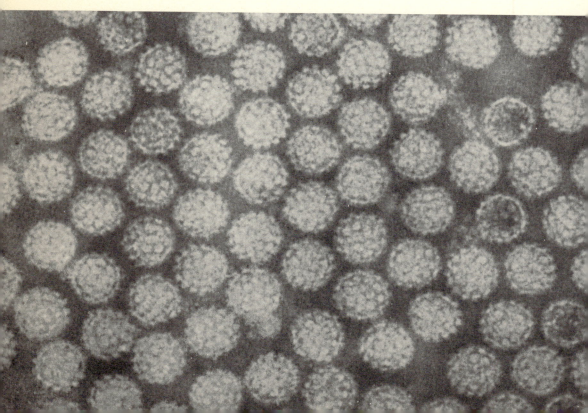

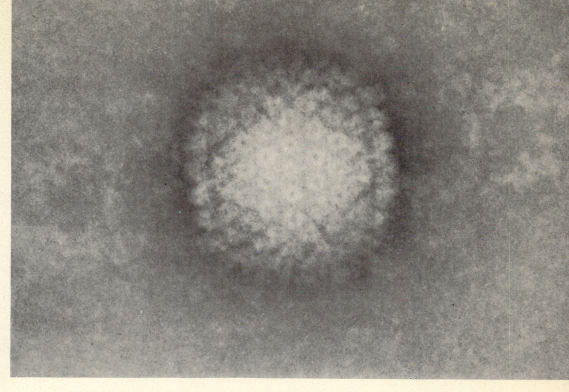

7. A herpes virus particle showing hollow prisms on the surface. One triangle face with six capsomeres along each edge can be made out.

8. Plate covered with a confluent growth of human cancer (HeLa) cells showing foci of cell destruction due to poliomyelitis virus.

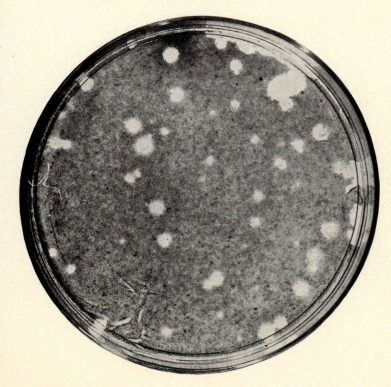

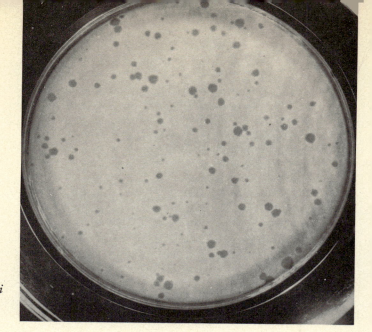

9. Plate covered with *Bacillus coli* showing clearings (plaques) produced by a bacteriophage.

10. Leaf of *Nicotiana glutinosa* manually inoculated with two forms of tobacco mosaic virus. Above (large lesions), tobacco form; below (small lesions), bean form.

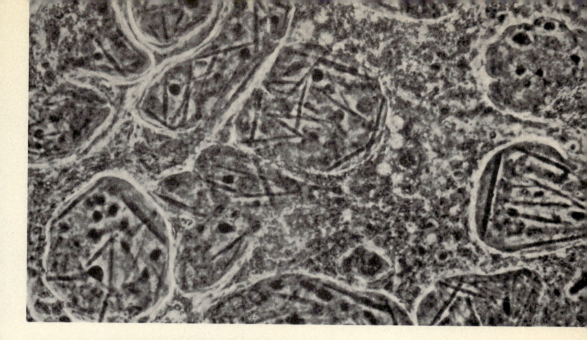

11. Cell in culture infected with an adenovirus. Elongated crystals of 'surplus protein' can be seen lying within nuclei.

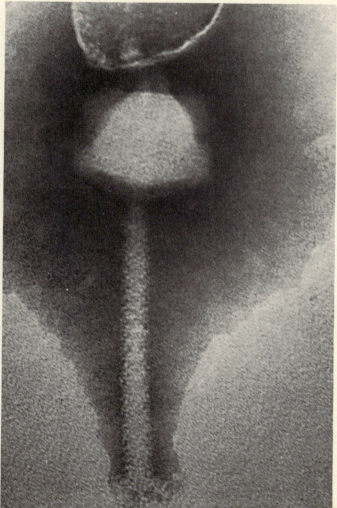

12. Phage showing long tail and hexagonal head.

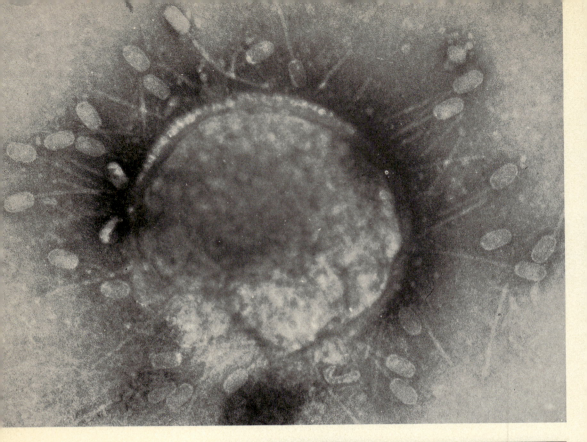

13. Bacteriophage particles being liberated by the bursting of an infected bacterium (a staphylococcus).

14. Electron-micrograph of a fowl red cell. The central white mass is the nucleus. Many small, round influenza-virus particles are shown adsorbed to the cell, and also long filaments, the filamentous form of the virus.

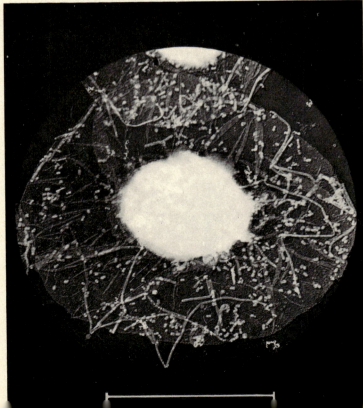

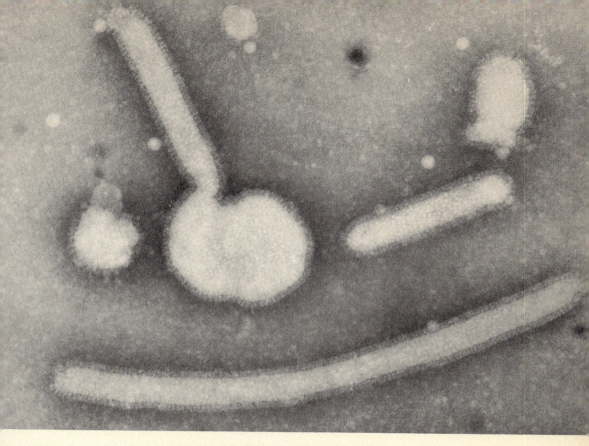

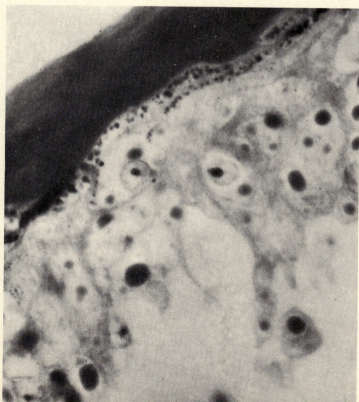

15. Filamentous and bizarre forms of influenza B virus. The surfaces are seen to be covered with spikes.

16. Section through the skin of a mouse infected with ectromelia (mouse-pox). The dark, rounded masses are inclusion-bodies, some of them representing colonies of virus.

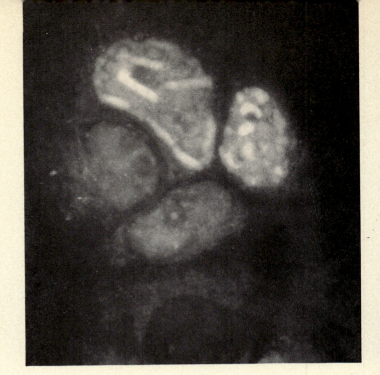

17. Cells infected with an adeno-virus and stained with fluorescent antibody. The white areas, in the nuclei, brightly shining under the microscope, indicate the where-abouts of the specific antigen.

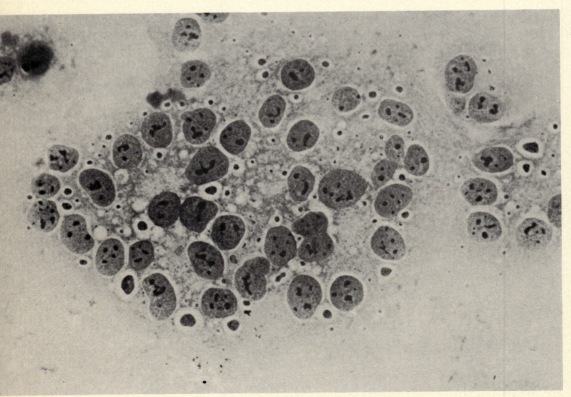

18. Syncytia, cells containing many nuclei, from cultures infected with respiratory syncytial virus; stained by haematoxylin and eosin.

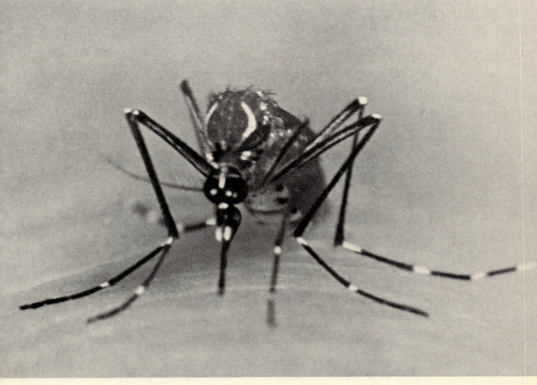

19 (Top) *Culicoides riethi*, female feeding.
20. *Aedes aegypti* female, in the act of feeding.

21. Tower in the Zika forest, Uganda, showing platforms erected at various levels for mosquito catching.

22. Heronries round a pig farm near Tokyo. Egrets in the trees are visible as tiny white dots.

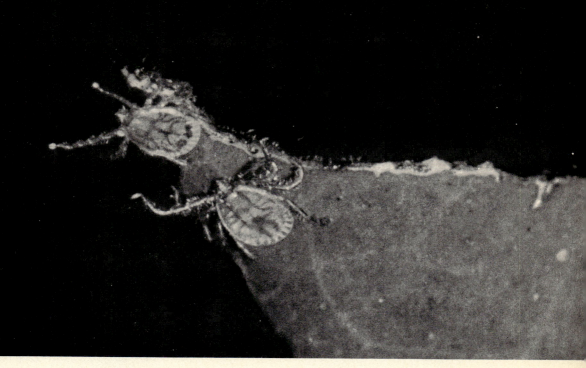

23. Adult *Haemaphysalis* ticks waiting to attach themselves to a passing animal. They have stretched out their legs in response to the presence of a man standing near.
24. Clusters of *Haemaphysalis spinigera* larvae on the underside of leaves in a forest in Mysore state, India.

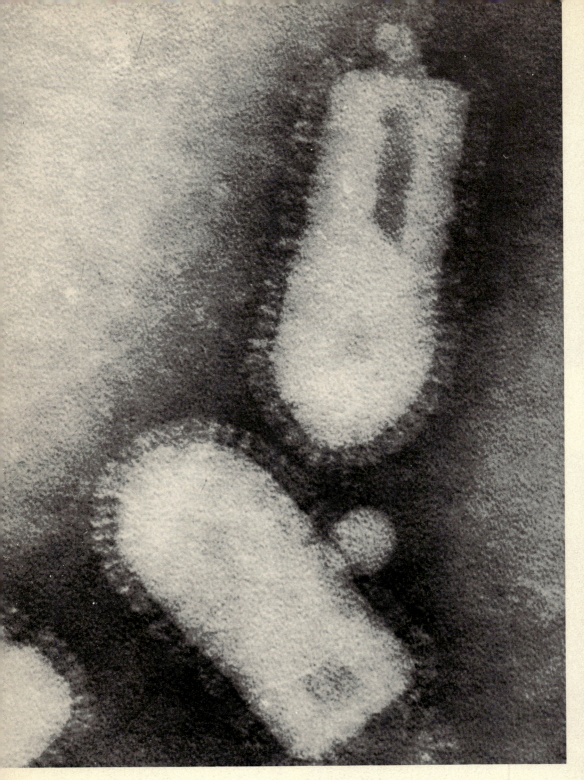

25. Electron-micrograph of vesicular stomatitis virus. It shows the characteristic shape, the surface spikes and a cavity inside.

26 (Opposite). Rod-shaped virus particles being liberated when a nuclear polyhedron from the gypsy moth (*Lynantria dispar*) is dissolved with weak alkali.

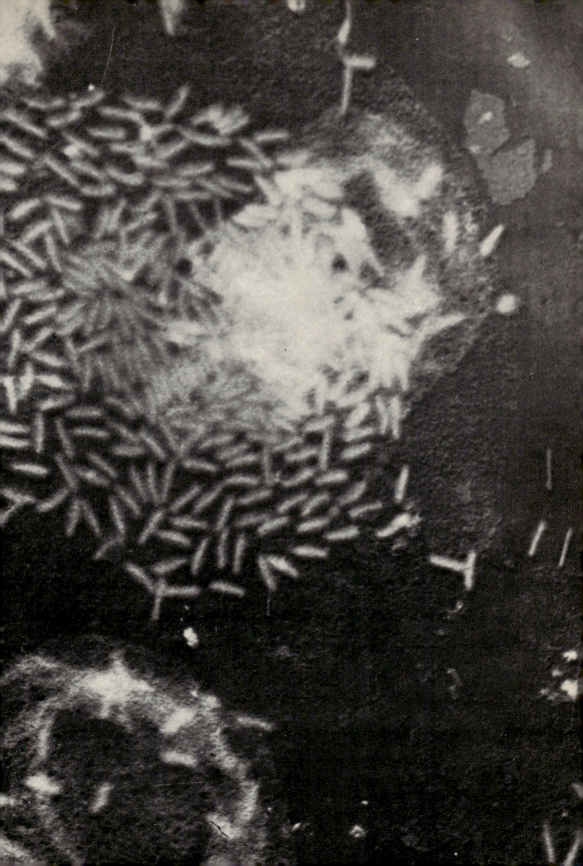

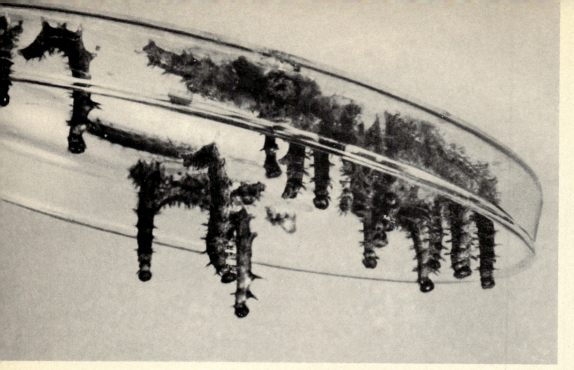

27. Larvae of tortoiseshell butterfly (*Aglais urticae*) dying from infection with a nuclear polyhedrosis.

28. Tortoiseshell butterfly larvae showing liquefaction as a result of polyhedrosis infection; its liquid contents are running downwards.

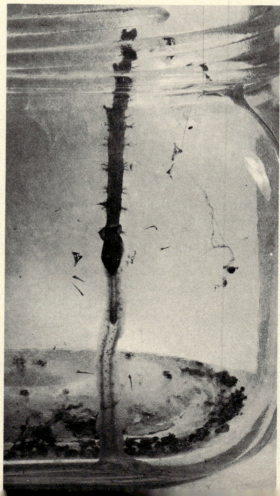

29. Tobacco plant systemically infected with a yellow form of cucumber mosaic virus.
30. Leaves of tomato systemically infected with the tomato aucuba strain of tobacco mosaic virus.

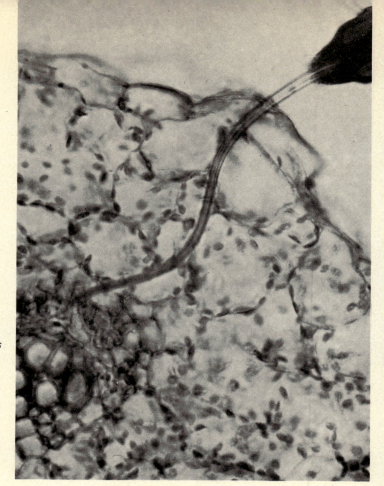

31. Stylet of the aphid *Myzus persicae* penetrating the phloem tissue of the vascular bundle of a sugar-beet leaf.

32. Pods of cacao. Above, from tree infected with cocoa swollen shoot virus; below, from normal tree.

33. Shoot of Arran Victory potato (right) grafted to tomato (left) symptomlessly infected with potato para-crinkle virus. The potato leaves show distortion and mosaic.

34. Cambridge Favourite strawberry plants. Left, normal; right, infected with arabis mosaic virus.

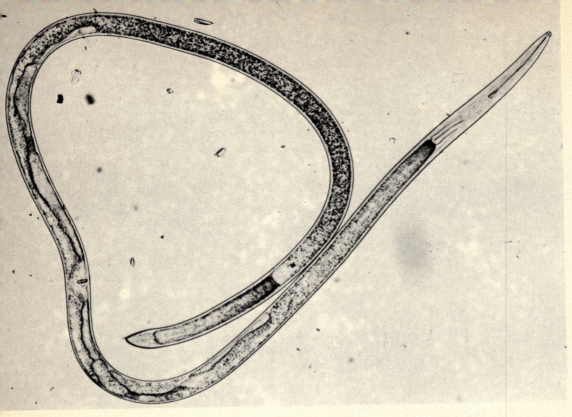

35. Adult female *Xiphinema diversicaudatum* (4–5 mm. long), a vector of arabis mosaic virus.

36. Zoosporangia of *Olpidium brassicae* in root cells of lettuce.

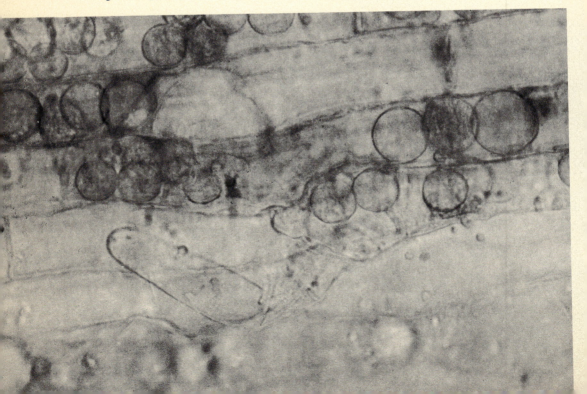

37. Views of Lullington, the upper photograph was taken in July, 1953, and the lower in July, 1961. The increase in scrub after myxomatosis had decimated the rabbit population is obvious.

38. A natural case of scrapie in a Swaledale sheep, photographed at the Institute of Research on Animal Diseases, Compton, Berkshire.

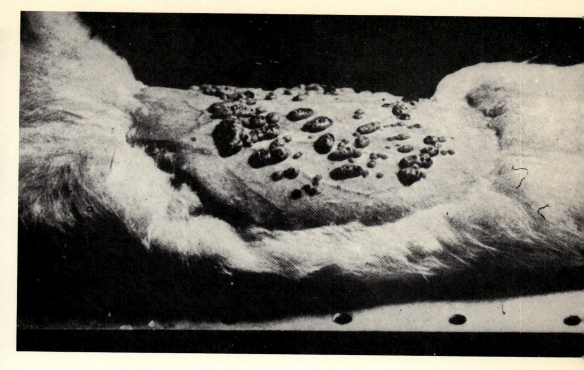

39. The development of warts after rubbing virus into the skin of a rabbit.
40. A rabbit in which warty growths on the flank have, at their lower edge, developed into an invasive cancer.

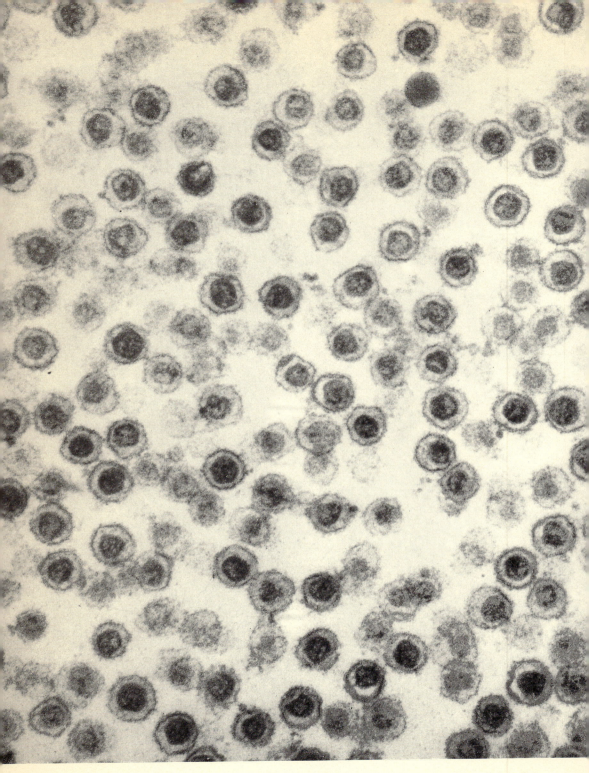

41. Electron-micrograph of a thin section of a pellet containing many particles of a leukaemia virus of mice.

considered more fully in chapter 18. Orf, or contagious pustular dermatitis, is thought to be carried largely by mechanical means. It may persist in the soil of an infected pasture for some months, waiting to make contact with the mouth of a grazing sheep or other species. For many of the other poxviruses the relative importance of the various available transmission-routes is in doubt; perhaps it varies.

The virus with the most astonishing spreading powers of all is that of foot-and-mouth disease. This is probably because it combines three useful properties: it is tough and not easily destroyed; it can get into the body very readily after contact with surfaces; and, most useful to it of all, only a tiny dose is necessary to infect. The virus is one of the smallest known, of the picornavirus family. Infection may occur by direct contact with an infected animal or as a result of encountering particles of virus dried on hay, on a person's boots or indeed almost any other object. Cattle are most affected, but sheep, goats and pigs are susceptible; also deer and, in Africa, impala and other antelopes. Most strains of virus are not very fatal but the loss of condition which follows the disease, especially in dairy cattle, often makes it not worth the farmers' while to keep the animals alive. The resulting economic loss in many countries runs into millions of pounds. Three sero-logical types of virus have been known for a long time, O, A and C and these are widespread; recently four more have been recognized, an Asiatic one and three from South Africa known as SAT 1, 2 and 3. In 1961 one of these South African ones turned up in Bahrein on the Arabian Gulf and soon spread to Iraq, Israel and Turkey. There were grave fears that it would spread all over Europe and energetic measures were taken to stop it. In particular, vaccination in a big way was carried out along the borders of Turkey with Greece and Bulgaria, using vaccine most of which was made in Britain. The spread was apparently successfully stopped. Similar ener-getic measures, combined with slaughter, were successful during a crisis in 1952 when there was a big outbreak in Mexico, threatening to spread to the United States.

A difficulty appears in that there arise variants of the seven types of virus; for instance the Middle East strains of the SAT 1 virus were just a little different from the previously known virus of that type. It therefore became necessary to make fresh vaccines against the new virus instead of using stocks of the original SAT 1. Although viruses which have been under study in the laboratory for a long time are remarkably stable, yet in the field the virus seems as regards its antigenic make-up to be disconcertingly labile.

The situation in the world today is this: Asia, Africa, most of continental Europe and South America are endemic areas. Here efforts are made to control the disease by vaccination and other measures. Australia, New Zealand and the United States are free from infection. By quarantine, combined with slaughter if necessary, they can hope to remain free. Britain is in a different position: the disease is not endemic but the virus is liable to be introduced in ways we are about to discuss. When it does get in, it has to be stamped out by slaughter, a painfully wasteful and expensive proceeding, though better for us as it happens than other policies available. There is often an outcry: 'Why can't we immunize as they do on the continent?' Such a policy would unfortunately mean that foot-and-mouth disease might well become endemic over here and would thereafter be always with us. For immunity to the disease following vaccination is only transient; as it wanes, cattle may have infections which are trivial and often unrecognized, but which may nevertheless spread the disease. So our best defence at present is to help infected countries to control their foot-and-mouth disease by trying to devise better vaccines, but to continue for ourselves to use the policy of slaughter.

There are at least three ways in which infection can reach Britain. Infected animals can bring it in; a policy of quarantine has, however, sufficed for some years to prevent that. We need, unlike many countries, to import meat, and much of this comes from South America. Here foot-and-mouth disease is endemic, and some gets through to us despite rigorous inspection and other precautions. If there were a little virus in our Sunday joint it would be well cooked, we hope, and would not in any case have much chance of infecting British livestock. The danger comes from offal and waste-products used to make swill for feeding pigs. In the muscle of lean meat the acidity which develops during *rigor mortis* kills off the virus; but in the blood, bones, lymph-nodes and other tissues not so much acid is found and these are the materials which find their way into pig-swill. Many primary outbreaks, that is those arising from direct introduction of virus, are in pigs. Strict regulations exist for the boiling of swill of this kind, but unfortunately they are not always rigorously carried out and it is then that trouble starts.

The third mode of entry of virus rests on circumstantial evidence and has never been proved to occur. While outbreaks caused by swill depend on what is happening in South America, other introductions take place at times when there are widespread outbreaks in countries across the channel and North Sea. Moreover they occur especially in counties directly facing the most seriously affected European countries. The theory is that virus is brought across by migrating birds, especially starlings. Recapture or shooting of ringed birds has given us considerable knowledge of the movements of birds, and some authorities hold that the facts fit in with the possibility that the virus enters Britain in such a way. The movements of starlings seem to fit particularly well and starlings are extremely abundant; moreover they like to hunt for food in pastures in the vicinity of cattle, which perhaps disturb insects for their benefit. Virus could undoubtedly survive on birds' feet during the time it takes them to cross the North Sea. Virus fed experimentally to starlings has survived passage right through their alimentary canals. Yet introduction of virus by starlings obviously only occurs rarely, so rarely that by the laws of chance it would be almost impossible to expect actually to detect virus on an immigrant starling. Though this theory has its adherents, it is fair to say that other authorities deny with equal fervour that it is at all probable. Another view is that seagulls are more likely culprits especially as they feed on garbage and have a habit of regurgitating food.

Scandinavian countries are relatively freer from trouble than are Germany, France and the Low Countries: people there have suspected that virus is carried across to Scandinavia by the prevailing winds. The virus is rather easily killed by light; yet one can suppose that it might get across on a nocturnal gale.

Foot-and-mouth disease readily infects guinea pigs and other small animals in the laboratory but these are not infectious for each other nor for cattle in contact with them. Hedgehogs, however, *are* infectious when they have the disease [104]. There are stories in the folk-lore that hedgehogs come and suck the milk of cows at night and make them go dry. It has been suggested that this story has its basis in cases of transmission of foot-and-mouth disease from hedgehogs to cows: that, however, is only a guess.

One could give many other instances in which transmission of viruses amongst animals was not simply by respiratory, enteric or blood-borne routes. Wart-viruses of man, rabbit, dog, cattle, horse and goat are all local infections transmitted from skin to skin by mechanical means. We will consider in this chapter only one more example, from one of the families of viruses pathogenic for insects [154]. These attack almost exclusively the larvae, particularly those of Lepidoptera (butterflies and moths), and the rather similar larvae of sawflies which belong to the Hymenoptera. A few are known infecting flies, lace-wing flies and members of other orders. Most of them are DNA-containing viruses; a few have nucleoprotein of the RNA type. Some dwell in the cytoplasm of cells, others in the nuclei. Among the latter is the big group of polyhedral viruses. These are so called because the rod-like virus particles are enclosed, sometimes in bundles, within a many-sided crystalline protein mass, the polyhedron (plate 26).

Caterpillars infected with one of these viruses feed normally until within a few days of death. Then they show changes in colour and seem to be rather swollen. At death they become completely flaccid and hang downwards in a characteristic way attached to a stem or leaf only by the posterior prolegs (plates 27 and 28). The skin has now become very fragile and is easily burst, liberating a milky fluid containing millions of virus-containing polyhedra. Infectious material can then be spread widely over the vegetation by the rain and the wind and is duly eaten by other caterpillars. Predatory insects and birds devouring the dead larvae may distribute infection farther afield and blowflies also frequent the rotting corpses of the dead larvae. A virus infecting the pine sawfly has been recovered from the faeces of a predatory bug, *Rhinocoris* and from those of the robin, *Erithacus*. The liquefied bodies of virus-infected larvae of the large white butterfly apparently attract healthy larvae which feed greedily on the bodies, only to perish themselves in due course. Some of the polyhedrosis viruses which live in cytoplasm of cells do not disrupt the larval skins in the manner described; virus is passed out in the faeces, but with similar results in the end. All these happenings lead to widespread epizootics and sometimes to the virtual annihilation of whole populations. This is a little surprising in so far as the infectivity of some polyhedra for larvae is not very great: it may take a few million polyhedra to infect a caterpillar. Nevertheless some success has been achieved in rearing the larvae of pests, infecting them with virus and spraying crops with watery suspensions of the dead larvae. By such means infestations of Canadian forests by the European pine sawfly (*Neodiprion sertifer*) have been brought under control. In another trial virus spraying was used against larvae of *Colias philodoce*, a butterfly related to the clouded yellow. There was already parasitic attack by the virus on the larvae, but the additional virus in the spray apparently increased the mortality and made the larvae die at a younger age before the crop had been seriously damaged. Twenty litres of virus suspension were used per acre and every litre contained five thousand million (10^9) polyhedra.

This is only one side of the story of insect viruses. They afford also one of the best illustrations of latent virus infection and its activation: chapter 20 will be dealing with that.

Chapter 15

Aphids, Hoppers and Beetles

IN THIS and the following chapter we shall leave the viruses of vertebrates to consider those infecting plants. Their numbers are enormous and more are being discovered all the time. One eminent plant virologist once told me that anyone with a little know-how could go out into his garden and discover a new plant-virus before breakfast. A survey carried out in Canada revealed that ten per cent of the wild plants sampled were carrying viruses, many of them of more than one kind. They are of very great economic importance and, not unnaturally, most of our knowledge concerns those viruses which affect crops useful to man. Plant-viruses are carried in a number of ways but transmission by insects, particularly aphids, is more important than is any other mechanism [26]. One estimate gives 150 viruses spread by aphids, forty by hoppers and related plant-bugs and fifteen by white flies. Grasshoppers, caterpillars, beetles, insects of other orders and mites were each responsible for transmitting only a few viruses.

Many plant-viruses can be transmitted mechanically, by inoculation or by merely rubbing the leaves of plants with sap. A method of estimating the amount of virus in a preparation is based on the counting of local lesions developing on leaves rubbed by a standard technique. Two differences from what is found with animal viruses may be mentioned. The amount of virus nucleoprotein in an infected plant may be very great, sometimes amounting to eighty per cent of the plant's total protein. On the other hand it may take a great deal of this material to produce any infection – for instance a million of the rod-like particles of tobacco-mosaic virus. The results of infecting a plant may be purely local or there may be a systemic infection involving a large part of the plant (plate 30). Many viruses which can be transmitted mechanically are not known to be carried by any insect vector; conversely many insect-transmitted ones will not infect when inoculated mechanically. One of the factors complicating these and other findings is that in ct and plant tissues may contain virus-inhibitors and unless these are somehow separated from the virus, infection will not occur. Finally, a virus may infect plants of widely separated families, and symptoms so produced may be altogether different. In fact, as with animal viruses, a virus producing inapparent infection in, let us say, a wild plant may cause serious trouble when carried to an introduced crop. There is no space to review any more of the immense mass of knowledge about the general aspects of plant virology; we must turn our attention to transmission by insects, and in the first place to the green-flies or aphids. The importance of these is very well known. Potato-growers in the south of Britain import their seed potatoes from Scotland just because aphids are fewer, later and less active in the colder climate farther north and so the seed is more likely to be free of aphid-borne viruses.

Aphids obtain their food by sucking plant-juices. They and other sucking insects are particularly effective transmitters of viruses because their sharp stylets readily penetrate plant cells. For plant cells have tougher, cellulose-containing, walls than animal viruses and so plant viruses can't just walk into a cell, as it were, and infect it: some damage to the cell must be done, so that it can get in. The life-history of aphids is all-important for their role as vectors. Many of them multiply on different hosts at different stages of their cycle and are thus apt to carry a virus perhaps from a weed to a cultivated crop. Then they have, as a rule, a number of generations of wingless insects or *apterae* produced asexually, followed by winged forms or *alatae*. Some viruses are transmitted better or worse according to whether apterae or alatae are concerned; obviously the winged forms can travel much farther and can spread a virus more widely. Factors such as overcrowding, changes in length of day, temperature and humidity are of importance in determining when a generation of alatae shall be produced; knowledge of such influences is therefore necessary to gain information about plant-virus ecology.

Transmission of viruses by aphids is of three kinds – stylet-borne, circulatory and propagative [108]; the last two overlap. Viruses may be stylet-borne by what are called non-persistent vectors. Some aphids acquire virus after quite short periods of feeding and are capable of transmitting almost immediately. On the other hand they may lose this power within a matter of hours. Even when this is not lost quite so quickly, it does not survive the next moult. Transmission seems to be merely mechanical, the virus taken up from the plant-cells being carried on the tips of the stylets. Aphids which have fasted for some time seem to be better able to transmit than others. Curiously enough it may happen that when the insects have fed for an hour or two they are unable to transmit virus as readily as when they have sucked only for a few minutes. It is not quite understood why, if virus is only carried mechanically, virus can be transmitted by some aphid species and not others. The structure of the tips of the stylets of some species may make them better able to retain virus; or it may be the depth of penetration of the stylets into the tissues which determines the result; or perhaps virus is inactivated by an inhibitor present in some species.

In the second type of transmission by aphids, the circulatory, it seems that virus has to be taken in, pass through the wall of the gut and so find its way to the salivary glands, ready to be given forth to the next plant which the insect probes for food. So it takes several days as a rule before an insect which has taken up virus is able to infect. On the other hand, these vectors are usually 'persistent' ones, able to transmit for the remainder of their lives. In many cases it is suspected that transmission is of the third, or propagative, sort, that is that the virus actually multiplies in the insect. This was proved to be the case for leaf-hoppers before it could be shown for aphids. Now, however, it has been claimed that potato-leaf roll viruses can multiply in the aphid *Myzus persicae*. Evidence was obtained that virus would multiply by passing it by injection from aphis to aphis in series. It must be added that some workers have failed to confirm this finding.

The circulatory and propagative methods of transmission are a good deal less frequent than the stylet-borne. Moreover, success seems to be variable. Some stocks or clones of an aphid species will transmit while others will fail to do so. With some aphids the spring winged form may function all right but not the apterae of summer.

The ecology of aphid-borne plant viruses is not unexpectedly found to be very closely tied to the ecology and behaviour of the aphids themselves [36]. All the known aphid vectors feed on more than one kind of plant, and so are well fitted to carry viruses from one plant species to another. Presence of certain weeds to serve as alternative hosts may be necessary to allow survival of an aphid species in a locality. On some plants aphids settle down to raise a family; on other plants they just take a meal and move on. In the latter case they may be more efficient vectors than in the former. Once they settle down and start to breed the wing muscles of the alatae tend to degenerate and so they do not move again and infect more plants. The casual feeders tend rather to pass from plant to plant. *Myzus persicae* is perhaps the best known of all aphid vectors and it is so important partly because of its notoriously restless disposition.

Agricultural practices, quite apart from the use of insecticides, affect aphid behaviour. Weeding may have a good effect by removing necessary alternative hosts or a bad one by getting rid of something which distracts attention from crops. The apterae, which have to walk instead of flying to a fresh host, naturally infect fewer plants than do alatae; but they do move about somewhat and when plants in a crop are closely spaced they will be more ready to do so. Apparently, like over-civilized men in towns, they hate to have to walk far. On the other hand, Storey found, in East Africa, that when Africans with rather poor methods of husbandry grew ground-nuts close together and did not weed too energetically the incidence of the disease 'rosette' was less than when theoretically better techniques were used. The virus was carried from outside by aphids which did not breed on the crops and the amount of disease, being related to the numbers of invad-ing aphids, was less serious when damage was spread over a greater number of plants.

Plants are more resistant to viruses as they grow older. Sugar-beet plants sown early in April were less troubled by yellow virus than those sown later, since they were older when the aphid attack came along. One has to con-sider also the predators and helpers of aphids. Swollen shoot of cacao in West Africa is one of the most economically damaging plant virus diseases known. It is carried by mealy-bugs or pseudococcids. Not only are seventeen species of these concerned but seventy-five species of ant, many of which help the mealy-bugs; three arachnids are also involved. A complex of over fifty species is thought to be involved in the case of a potato aphid, there be-ing predators, parasites and hyperparasites (parasites of parasites) concern-ed. In Britain overwintering of aphids and their predators is easier than farther north. Spring populations of certain aphids are therefore greater but the build-up of predators and parasites is faster, so there are fewer aphids in summer.

Where aphids bring in virus from weeds at the periphery of a field, inci-dence of a disease may well be chiefly in a zone at the edge. But aphid move-ments are not only on a local basis; the insects may be carried for very long distances by the wind and at heights of several thousand feet. Much study has been given to the times of their flights and the conditions of tempera-ture and humidity which determine their movements. Methods of estimat-ing aphid populations have been worked out, and on the basis of all the relevant data, entomologists are able to issue forecasts of the dangers to be expected from particular aphid-invasions. These may tell the farmer that spraying is urgently needed or, on the other hand, that he may give it a miss this year and save his money.

This brief account will, I hope, make it clear what a large number of factors have to be taken into account in working out the ecology of aphid-borne plant-viruses.

The leaf-hoppers or cicadellids form the next group of vectors to be con-sidered. They are mainly persistent vectors and mechanical transmission is usually not demonstrable. They are of special interest since it was with some of them that multiplication in the insect vector was first demonstrated. These are in fact not just plant-viruses, but plant-insect viruses. The rice stunt virus passes through the eggs of an infected leaf-hopper to its progeny, though it does not go through the sperm. A Japanese worker, Fukushi, passed the virus through six generations of hoppers under conditions such that they could not have picked up virus from the plants they fed on. Black [22] carried out similar experiments with the virus of clover club leaf, passing it through twenty-one generations of leaf-hoppers over a period of five years. All this time the hoppers were fed on a variety of lucerne (or alfalfa) immune to the virus. Aster yellows virus was passed by Maramarosch [107] ten times in series by infection from hopper to hopper. Another virus, that of wound tumour of sweet clover (*Melilotus*) has been similarly passed; this virus is morphologically like an animal virus, reovirus, and, like it, contains RNA which, unlike the RNA of viruses in general, is double-stranded [64]. On the other hand, the virus of curly-top of beet seems not to multiply; it shows a minimum extrinsic incubation period in its leaf-hopper vectors of only four hours, though their maximum ability to transmit may not be attained till after two days of feeding. The transmitting power waned over a period of some weeks. Viruses may pass directly to the salivary glands of a hopper without multiplying, or they may multiply in some tissue first. Storey [161] found that the maize streak virus could be transmitted by some races of the hopper *Cicadulina mbila,* but not by others. Apparently this was because in some hoppers it could not penetrate the gut wall. So he punctured the mid-gut of the insects and let the virus through, and they were then able to transmit. The existence of these plant-insect viruses raises an interesting question concerning virus evolution. They are mostly highly specific as regards their insect-vectors, but can infect a wide range of plants. This suggests that they are better adapted to their insect hosts than to the plants and may, in fact, have originated as insect viruses, becoming secondarily adapted to multiply in plants. This view is supported by the fact they produce more disease in plants than in insects, though, as we shall see, not all of them are wholly harmless for their vectors.

Many of the factors considered in the case of aphid-borne viruses apply also to viruses of this group. Hoppers have no wingless adult phase and are, in general, more active insects than aphids: they also may be carried for long distances by winds. In the western USA large populations of the beet leaf-hopper build up on summer or winter annual weeds; these may be in quite different areas. Wind may carry the hoppers for long distances in the spring to breed in widely separated districts. Presence of appropriate weed-hosts may be very necessary to enable insects to survive in the absence of their ordinary host-plants; if no suitable weeds are available the hoppers have to migrate.

Some other sucking insects carry and transmit plant viruses. Tomato spotted wilt is the only virus known to be carried by thrips. The virus is persistent in the vector. For one species, only the larvae can acquire virus though it can be subsequently transmitted after they have reached the adult stage. In India, whiteflies (Aleyrodidae) are probably the most important transmitters of plant viruses. Infectious chlorosis of *Abutilon* was a puzzle for a long time to workers in Europe and North America, where it could only be transmitted by grafting. Then it was found that in Brazil, a whitefly vector was present and the ecological puzzle was cleared up [150].

Mealy-bugs (Coccidae) have already been mentioned as transmitters of swollen shoot of cacao (plate 32). The virus infects certain forest trees in nature and spreads from them to the cacao trees. A few viruses are carried by mites in which they persist for some days, though infection is not passed through the egg.

Biting insects are much less concerned than are the sucking ones in virus transmission. They have not, however, been sought for nearly so diligently. It would not, perhaps, be expected that biters would be very efficient vectors, for they cannot damage many cells sufficiently to permit virus entry without also destroying them. In fact the biters mostly carry the viruses which can also be transmitted mechanically, though rather unexpectedly virus may persist on their mouth-parts for some time. Flea-beetles are perhaps the most important among the biters, being the vectors of turnip yellow mosaic and turnip crinkle viruses. They do not infect as aphids and hoppers do, having no salivary glands: it has been suggested that they may regurgitate infective juice from their fore-guts in the course of feeding. Grasshoppers remain infective for some hours after feeding on plants with tobacco-mosaic and some other virus infections. Larvae of moths have also been shown to carry a few viruses; so have leaf-mining flies, leaving virus on their ovipositors.

The last matter to be considered in this chapter is the effect of plant viruses on their insect vectors. The mosquitoes and ticks which carry animal viruses are, so far as we know, affected by the viruses neither for good nor ill; and the same is probably true for most of the insects which carry plant viruses. Not, however, for all. Maramorosch [109] has summarized the available evidence on 'Harmful and beneficial effects of plant viruses in insects'. The first effects on insect vectors of plant viruses to be described were those detectable by microscopical examination of insect tissues. Bodies within cells were found in hoppers of the genus *Delphacodes* infected with oat mosaic or winter wheat mosaic. In the affected plants there were intracellular bodies either in the form of needle-like crystals or having a semi-liquid appearance, and similar things were found in the insects. Hoppers infected with aster yellows also showed microscopical changes; in particular the nuclei of many cells appeared star-shaped instead of round. Nevertheless the infected hoppers seemed to breed rather better than normally. Changes have also been detected in the tissues of virus-carrying aphids. Particles believed to represent virus have been seen by electron-microscopy in sections of tissues of infected insects. Biochemical studies have also detected changes; the metabolism of insects may change when they take up virus – in particular oxygen consumption may be reduced.

These microscopical and biochemical changes are interesting enough, but an adverse effect on the health and life of the insects would be much more dramatic and convincing and this has now been demonstrated. The western X-disease of stone-fruits, which also causes a yellows disease in celery, greatly reduces the life-span of its hopper vector, *Colladonus montanus*; infected hoppers survived on the average thirty days instead of sixty-five. The real difference was probably still greater, as not every hopper which fed on an infected plant managed to acquire virus. Moreover the hoppers' fecundity fell; only one-third as many offspring were produced. This, of course, may have been due in part to the fact that the females did not live as long.

The rice dwarf virus, the first virus proved to be insect-borne, has been particularly studied in Japan. It can be transmitted from generation to generation in the egg of its normal vector, the green rice hopper, *Nephotettix*. It nevertheless causes considerable mortality among the hopper nymphs, especially soon after hatching. Nymphs which acquire the virus by sucking infected rice, instead of congenitally, are also apt to die but not so early, most frequently after their fourth moult.

In contrast to all this, a virus may actually prove beneficial to an infected insect. Severin [143] found that the nymphs of nine leaf-hopper species developed well on plants of celery or aster infected with the Californian aster yellows virus, whereas they soon died if put on to healthy plants. The aphid, *Aphis fabae*, produced more offspring than normal when on sugar beet plants with mosaic virus. Maramorosch [109] discovered a remarkable phenomenon in the maize hopper, *Dalbulus maidis*. This insect normally feeds only on maize and a related grass; on normal asters it dies. It can acquire aster yellows virus but not transmit it. When, however, it is fed on plants already infected with aster yellows it flourishes and, moreover, after

131

some time is able to flourish also when transferred to uninfected asters, to carrots or to rye. The effect appears to be on the power of the hopper to digest foods which were formerly wholly indigestible. It happens that the aster yellows virus can be killed by moderate heat. So some aster-yellows infected hoppers were cured of their infection by holding for eight days at 36°C. When this was done they were no longer able to survive on normal asters and digest their juices.

Maramorosch raises the question of whether we may not be overlooking something of wider importance in virus ecology. We assume that virus infections will do us harm, or at the best be tolerated. Is this necessarily true? If there were a lot of viruses in the world which were actually beneficial, should we ever recognize their existence?

Chapter 16

Boots, Worms and Fungi

THE WORDS at the head of this chapter refer to some of the agents concerned in the transmission of plant viruses otherwise than with the help of insects.

At one time workers knew of very many plant-diseases of which a virus causation was suspected, yet no transmission could be achieved except by means of grafts. For many of them, other routes of transmission have since been discovered, for instance the disease of *Abutilon* mentioned in the last chapter, a puzzle until a Brazilian white-fly was brought into the picture [150]. There remain, however, a number of well-known viruses for which grafting seems to be the only means of transfer. Best known of these is para-crinkle, a virus present in all stocks of King Edward potatoes (plate 33); one well-studied strain of this causes no disease in them, but does so when grafts of King Edward are made into other kinds of potato [16]. There was for some time an argument as to whether the active agent might not be a normal constituent of King Edwards, pathogenic in other varieties. This is no longer believed. For one thing King Edward plants raised from true seed – not 'seed potatoes' in the usual sense – do not carry the para-crinkle virus; for another, many strains of the virus are now known to be aphid-transmitted. What is believed to have happened is that an insect-borne virus has been carried on in potatoes propagated only vegetatively for year after year and has finally lost the ability to be carried by insects.

There are other viruses, including that of tobacco mosaic, which are not transmitted by insects but are very readily transferred to other plants mechanically [185]. This makes them particularly easy to study in the laboratory in a quantitative manner, and a lot of work has been carried out on the conditions which make such transfer more efficient. For example it proves to be useful to incorporate a little carborundum in virus fluids rubbed on to leaves: this produces just that little extra damage which helps virus to 'take'. The presence of phosphate may also prove useful. Other factors affecting the result are temperature and light. Inoculation of virus from plant to plant by syringe has not proved to be helpful; in fact it may fail even with leaf-hopper viruses which can be transferred from insect to insect by this means.

Viruses which are mechanically transmissible in nature are mostly highly resistant ones present in plants in great quantity. Tobacco mosaic (plate 29) can be spread widely by means of the boots of agricultural workers, farm machinery and implements or may even come from the workers' cigarette ends, the tobacco of which was made from virus-infected plants. The route of a contaminated tractor may show up as a ribbon of diseased plants across a field. Virus may be also transferred from plant to plant when these are in close contact so that leaves touch and brush each other. Wheat mosaic virus is said to be wind-borne for considerable distances, and tobacco necrosis virus may be either air- or water-borne. Dust storms have in some cases been blamed for spreading viruses. It may be that some of these viruses were originally insect-borne but, acquiring high resistance to inactivation, were able to spread by other means and so could dispense with the need to invoke insect help.

A more important class of viruses comprises those which are soil-borne. Some of these are very important economically. Wheat mosaic virus is estimated to have decreased the 1957 wheat crop in Kansas by four million dollars. Contact between roots may spread some viruses and others seem to be carried when spontaneous root-grafting occurs; and yet others, such as cucumber mosaic, by root-feeding aphids. Harrison [73], however, restricts the term 'soil-borne viruses' to those with an underground natural method of spread which does not depend simply on contact between tissues of infected and healthy plants. Simple diffusion through the soil does not seem able to account for everything; some active agent must be concerned in many instances. Often none such has been identified, but virus-transmission may fail if virus is simply added to sterilized soil and plants are grown in that. Indeed there are no claims to have successfully transmitted the viruses under discussion in sterile soil. Mechanically transmissible viruses such as tobacco mosaic may, however, persist in soil for long periods of time and remain infectious.

Soil-borne viruses may remain in the roots of the plant, and in such cases there may be no evidence of infection; alternatively, there may be some stunting of a plant's growth or underground symptoms such as corky arcs in potatoes or necrotic spots in hyacinth bulbs. Virus may, however, pass up to produce systemic infection of the rest of the plant and various symptoms may then develop. It is of interest that many soil-borne viruses develop local races, a little different from those found elsewhere; this could be a reflection of the fact that they do not readily spread widely and thus show the effects of geographical isolation such as are familiar to zoologists and botanists. On the other hand differences between races may turn out to correspond to transmission by different species of vector.

The agents which have been definitely incriminated in underground transmission of viruses are eelworms (nematodes) and fungi. An important disease in vineyards is due to grape vine fanleaf virus, which is also known by several other names. It is carried by the nematode *Xiphinema index*, described as 'a large migratory dagger-worm with a long mouth-spear' (plate 35). To allay alarm among addicts of science-fiction it should be added that 'large' in this connection means about 4 mm. long. A worm of another vector species (*Trichodorus*) is described as puncturing a cell in a plant-root 'by a rapid probing motion of the stylet; the contents of the cell were then withdrawn. After one to four and a half minutes the nematodes moved to another cell, which sometimes was punctured only ten seconds after leaving the first. One nematode was observed feeding in this manner for eight hours'. Females and larvae of *X. index* carry fan-leaf virus to grapes: the virus may persist in the worm up to eight months, but whether or not it survives a moult in the larva is not known. There is no evidence that it passes through the egg and whether or not it actually multiplies in the worms has not been really looked into. It may be mentioned, however, that at least one instance of multiplication of a virus in a nematode is on record, though it was not a plant-parasitic one. Nematodes can pick up virus from a plant in the way described but apparently cannot do so directly from soil.

Nematode-borne viruses are often carried by only one worm species, but they may affect a number of different plants causing different symptoms. They have thus come to be called by a variety of names and it has taken time to disprove the idea that a whole lot of different viruses were involved. Thus four viruses related to ring-spot of raspberry have been known, between them, by fifteen different names. These are viruses of polyhedral form; many of them may be carried in the seeds of the host-plants and this infection may come via either ovule or pollen.

Other nematode-borne viruses are of tubular shape; the most important among them is tobacco-rattle. The infectivity of this virus is very variable and infected plants may contain a lot of infective virus nucleic acid, only a part of which ever gets built into the stable rod-shaped virus particles. The question has been raised of whether nucleic acid may be able to carry infection in nature without need for incorporation into complete virus.

The ecology of nematode-borne viruses is naturally closely related to the ecology of the eelworms themselves. They, and accordingly their associated viruses, are some of them prevalent in light loam or light peaty soil, others in light shingle, clay or in peaty fen. Their occurrence in soil may be quite local, but they may be found down to a depth of two feet. The viruses concerned are fairly stable; they have to be, for the worms only feed intermittently. They may persist in the worms themselves, in surviving roots of plants or in dormant seed. *Xiphinema diversicaudatum*, which carries Arabis mosaic, an important pathogen of strawberries (plate 34), tends to persist mainly in woody plants [74]. These, near strawberry crops, are mostly in hedges and so most damage to the strawberries is near the hedges. They do not spread fast; fanleaf of grapevines may spread less than two feet a year, while sometimes raspberry ringspot persists in local patches and does not seem to spread at all. Obviously, moving plants with a ball of earth round the roots may have the effect of moving eelworms and viruses at the same time. However, persistence in soil is not for ever, and some viruses can be thwarted by a variety of treatments; alternation of crops will only be effective if the virus to be controlled is not one with a wide host-range.

Another important way in which soil-borne plant viruses are transmitted is with the help of fungi [92]. These fungi belong to the genus *Olpidium* and have the following life-history. Structures called zoosporangia in cells at the root-surface of an infected plant liberate zoospores (plate 36), which are small round bodies able to move around by lashing a cilium or tail-like thread several times longer than the bodies themselves. These zoospores finally reach another root-cell and after entering it give rise to another zoosporangium, and so it goes on. Tobacco necrosis virus (plate 30) is able to infect lettuce and certain beans in the presence of the zoospores, presumably entering the cell along with them. Exposure of roots to virus with zoospores present for one minute was enough to permit infection to occur. This infection took place more readily when the virus had access to the roots along with the zoospores than when it was added later; some, however, managed to get in when it was added as much as four hours afterwards. There was no evidence to suggest that the virus was actually inside the fungus's zoospores. These managed to carry virus into the cells so long as they penetrated it; it wasn't necessary for them to continue to develop themselves.

Mechanical infection of roots with the tobacco necrosis virus was possible, but it was very inefficient when compared with what happened when zoospores were present. Incidentally, tobacco necrosis virus is of interest from another point of view. It is sometimes accompanied by a 'satellite virus', one which is quite distinct from it but is only able to infect cells which are also infected with tobacco necrosis. It is likely that two other viruses, those of tobacco stunt and lettuce big-vein are also transmitted with the help of *Olpidium* fungi.

Mention may be made of one other means by which plant viruses are spread – by means of dodders. In Britain there is only one common member of this family of parasitic plants; this is seen in the form of a network of red threads over gorse and heather plants, with tiny pink flowers. Elsewhere, other dodders attack various other plants and some viruses make use of them to spread to fresh victims.

Part 3
Other Aspects of Ecology

Chapter 17

Rabies and more about Reservoirs

RABIES OR HYDROPHOBIA in its most familiar form is a terrifying fatal disease which may be transmitted to human beings by the bite of a mad dog or cat. The picture from the point of view of virus ecology is an unusual one. The virus enters at the site of the bite of the rabid animal and passes along the nerves to reach the brain, where it gives rise to encephalitis. The effects of the encephalitis are two-fold, to produce the violent excitement of 'furious rabies' or the apathy and paralysis of 'dumb rabies'. Both effects are usually present but one is liable to predominate over the other. Animals with the furious form of the disease tend to snap at strangers and even inanimate objects and later fail to recognize their owners and bite them also. The bite is liable to transmit infection because the virus passes not only to the brain but also to the salivary glands, so that saliva is often highly infectious; indeed a rabid dog may infect by licking, the virus gaining entry through some trivial scratch. Another effect of the encephalitis is to make swallowing difficult; the painful spasms associated with efforts to do this in man cause the fear of water which gives rabies its other name, hydrophobia. In dogs difficulty in swallowing leads to drooling, and saliva is thus very freely available to infect. Thus in rabies it is the symptoms of severe disease which themselves assist its transmission to other victims.

In the urban areas of many countries in the world dogs and cats are kept so commonly that the disease is readily kept going in the manner described. It is readily spread to new areas in that mad dogs tend to run for considerable distances. Some fortunate countries including Britain and Australia manage to keep free from rabies by imposing a six-months quarantine on dogs coming into the country; a long period is necessary since the incubation period may be long. There is a periodical outcry from pet-lovers against this quarantine, but anyone with personal knowledge of the terrible effects of the disease should regard some inconvenience as well worth while to keep this pest out of the country. It is true that dogs can be vaccinated, but it is doubtful if any scheme of vaccination against any virus is a hundred per cent perfect.

Rabies in mad dogs, however, is by no means the whole story. We saw in chapter 11 how the urban cycle of yellow fever in man and *Aëdes aegypti* was something superimposed on a cycle in the jungle tree-tops. We see something similar in rabies: the disease exists as 'sylvatic rabies' in many wild animals, and the reservoir of the disease is different in different parts of the world. In parts of Europe rabies was formerly endemic and these areas served as starting-points for migrating epizootics amongst wolves and foxes. Reference was made earlier to travelling outbreaks of yellow fever amongst wild monkeys in South America. In both instances the epizootics had to travel because the supply of susceptible animals was exhausted locally. When dogs once became infected the future of the rabies virus seemed assured, since opportunities for its spread were much better.

Rabies is particularly common in India. In Madras state alone, 1,475 human cases were reported in 1941; and it should be mentioned that by no means every person bitten by a rabid dog will contract the disease. Wild dogs, foxes and jackals were the common sources of infection, though of course most human infections come from domestic dogs. It is likely that a more important reservoir is in the carnivores of the family Viverridae which includes the mongooses and civet-cats. H. N. Johnson [89] has advanced reasons for believing that animals of this family and of the Mustelidae (weasels and skunks) may be of great importance in the ecology of rabies; Australia and Hawaii are among the few places in the world where these families are not represented and in them there is no rabies. In Iran, rabies infections have come particularly from wolves, and because wolf-bites are usually severe, mortality from wolf-rabies has been very high.

In Kenya there have been epidemics of rabies among jackals. A disease affecting dogs in West Africa known as oulou-fato was for long not recognized as rabies; it exists as the dumb form and seems not to be established in dogs, but to exist in some reservoir as yet unidentified. In parts of South Africa yellow mongooses and related species are important vectors. They are especially dangerous in that when they have the disease they are apt to become unusually tame, so that children may play with them and receive a nip or contamination by their saliva. In the areas of mongoose-rabies, dogs seem not to be affected.

Rabies exists also in the Arctic. Here it causes outbreaks among arctic foxes but may be more truly endemic in the ermine. During epidemics it may spread to wolves and sled-dogs. The larger outbreaks seem to follow the periods of population-explosions among lemmings and voles. These are followed by increases in their carnivorous predators, and finally there is an increase in the rabies virus which preys on the carnivores.

In North America sylvatic rabies exists particularly in skunks and foxes.

Rabies in dogs has now been largely controlled by vaccination, so the reservoir which remains in wild animals is all the more evident; it may even be increasing. There have certainly been epidemic waves among wild animals and when these occur there has been a spill-over into dogs. There are many stories concerning skunk rabies in the last century. Skunks are reported as 'charging' into camps of buffalo hunters, 'snapping at everything, their small teeth sharp and deadly'. The small spotted skunk was particularly concerned and because they were recognized as vectors of hydrophobia they were known locally as 'phobey cats'. Recently more blame has been allotted to the larger, striped skunk, which lives nearer human habitations. The map (figure 17) taken from a review by E. S. Tierkel [169] shows that fox rabies and skunk rabies occur in rather different areas of the USA, fox rabies in the east and south, skunk rabies in the mid-west and in California, with some overlap. In Puerto Rico an introduced species of mongoose is the main host.

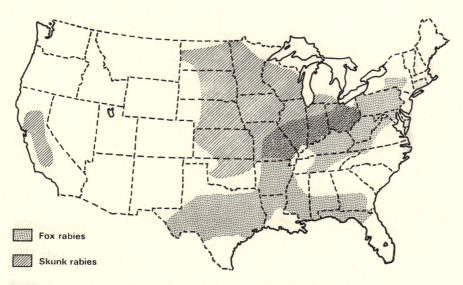

Fox rabies

Skunk rabies

Figure 17. Map showing distribution of fox and skunk rabies.

Much interest has centred in recent years on the role of bats [54]. Between 1929 and 1935, eighty-nine people in Trinidad died of an ascending paralysis, at first thought to be due to poliomyelitis; a paralytic disease was simultaneously occurring among cattle and this was diagnosed as botulism. Drs Hurst and Pawan [86] showed that the infection was in fact rabies and that it was carried by vampire bats. These bats, *Desmodus rotundus*, live on blood. They attack their victims at night, making crater-like holes with their sharp incisors and lapping up the blood which escapes. Stories are told of such attacks on a big toe left protruding from under the bed-clothes. For some years it had been known that rabies was transmitted by vampires to cattle in South America, and in 1956 a million head of livestock are estimated to have been lost. Rabies in cattle is fortunately almost always of the dumb or paralytic type; infected beasts do not go around biting and so spreading infection. We have here, in fact, another blind-alley infection. The vampires themselves may or may not show symptoms. A vampire has been shown to be capable of infecting as a symptomless carrier over a period of five months. On the other hand at times of much rabies prevalence vampires have been seen by ranchers behaving abnormally, flying in the daytime and fighting with each other. The problem of controlling them is a difficult one, especially as they have become more numerous; increases in numbers of cattle have provided them with very readily available sources of food.

Vampire bats do not spread north of the Mexican border, but it has been discovered since 1953 that insectivorous bats also may harbour rabies in the USA. Several cases of rabies have occurred in human beings who have picked up and been bitten by sick bats, or who have been actually attacked by these bats. The bats concerned have been chiefly the Mexican free-tailed bat (*Tadarida*) and migratory bats of the genera *Dasypterus* and *Lasimus*. Fruit-eating bats have also been found to be infected. The *Tadarida* are colonial cave-dwelling bats which could perhaps readily transmit infection in the course of a little friendly scuffling and nipping in their day-time roosts. Infection may also be by the respiratory route. Foxes and coyotes confined in cages in caves where there was bat rabies have become infected; inhalation of virus seems to have been responsible. Rabies has been recovered also from insectivorous bats in Jugoslavia and Turkey. Fortunately bats do not migrate to Britain across the Channel and there is thus no danger that they could smuggle rabies virus into the country, evading our quarantine. How far the insectivorous bats can act as symptomless carriers, as vampires do, is unsettled. Virus has been recovered from 14·6 per cent of 199 individually tested bats collected in flight; it was present in some in salivary glands but not in brain. The infected bats may have been in the incubation-stage of rabies but it seems likely that some at least were carriers. One bat, kept alone in captivity, yielded virus in its saliva sixteen months after capture [169]. A recent discovery is that rabies virus is present in five per cent of apparently normal bats in the state of New Jersey; this is particularly surprising since no case of rabies in man or dogs has been recorded in that state for nearly ten years. There is evidence that in bats the virus may persist during hibernation in the brown fat, a collection of curious tissue in the small of the back, having a role in the physiology of hibernation.

Discovery that vampires, and perhaps other bats, could 'carry' rabies naturally raised the question of whether other animals could do so too. If not it could be argued that there is evidence here of at least a partial host-parasite equilibrium, suggesting that bats might be the original hosts of the virus. Johnson [89], however, thinks that bat rabies is 'an aberrant cycle of the disease, part of the recent general involvement of wild life with rabies'. It seems to him more likely that the Viverridae and Mustelidae, the mongooses and weasels, are the real reservoirs. Weasels and stoats or ermines are good climbers and could readily prey on bats and so be responsible for introducing infection among them. It used to be supposed that rabies was a fatal disease in all species. It now appears that this is untrue, not only for bats, but for rats, mice and other species. A recent survey organized at the Communicable Disease Centre in Atlanta, Georgia, sought light on the subject by looking for antibodies in the sera of various wild mammals. Antibodies were detected in the blood of 4·5 per cent of foxes, 5·6 per cent raccoons, 1·8 per cent opossums, 18·5 per cent bobcats, 14·5 per cent skunks. These sera all came from areas with a recent history of outbreaks of fox rabies; tests on sera from three hundred animals from rabies-free areas gave wholly negative results. The positive results could of course simply mean that these animals had had an infection from which they had recovered; the question of possible symptomless carriage of virus by these species remains unsettled. It seems unlikely that foxes can act as carriers for in naturally infected foxes salivary glands never contained virus unless the central nervous system did also. Foxes should, however, be efficient spreaders of virus, for rabies does not kill them quickly and one animal had infectious saliva over a period of seventeen days. There was no evidence in these studies suggesting that small rodents could serve as reservoirs.

Recently Johnson [89] has made a suggestion which introduces a new idea into virus ecology. He proceeds from the idea that mustelids may be the real reservoir. The virus infecting skunks and allied species inapparently might well spread by the respiratory route or, because of its affinity for kidney and mammary glands, through the urine or milk. Presumably only a small proportion of the animals would suffer from fatal infections. The mustelids, foxes and other carnivores will do very well at times of rodent plagues; but such plagues are commonly followed by sudden crashes in numbers of rodents. The carnivora, whose numbers have also increased, then face a dangerous shortage of food. During the times of increase of prey and then of predators, rabies has spread from the mustelids to the foxes and larger carnivores. These now die in large numbers: coyotes have been almost exterminated locally as a result of rabies-outbreaks. The mustelids remain relatively untouched, because more resistant, and joyfully see their main competitors eliminated from the scene. If these suggestions are sound we can see how a latent virus-infection may actually help a vertebrate to survive. A puzzle remains: we shall see in chapter 18 how in the face of very lethal myxomatosis, Australian rabbits rapidly developed a considerable degree of resistance. Why haven't the foxes and wolves evolved rabies-resistant strains? We may consider two situations. First we have cycles in foxes and wolves which may not continue indefinitely; infection in these species has to be renewed from time to time from the reservoir in mustelids, so the evolutionary pressure would be only intermittent and so relatively ineffective. Second, there may be in some parts of the world an indefinitely continuing cycle in dogs. The virus's survival would then depend on its ability to excite its victims so that more dogs were bitten. If then the dogs became resistant, there would be pressure on the virus to become more aggressive, so preserving the *status quo*. A rather parallel situation will be considered in the next chapter (see p. 155).

It must be added that Johnson's views on the importance of mustelids are by no means everywhere accepted. Others think that bats have a better claim to be considered as the fundamental reservoir host.

The virus of rabies is, on morphological grounds, closely related to the myxoviruses [3]. Another disease has been confused with it because of the violent excitement in its victims; this is called pseudorabies, but it is quite unrelated to rabies itself, being a member of the Herpesvirus group. It is also known as Aujeszky's disease after the discoverer and, in America, as mad itch. This is because the excitement referred to is due to intolerable itching, such that affected animals seriously damage themselves by biting or scratching at the skin, usually of the head and hind-quarters. Later, fits and paralysis are followed by death. Pseudorabies attacks especially cattle, but also sheep, cats, dogs and mink; in all these animals it is invariably fatal. A few human infections are recorded; but the illness was not very severe.

The key to its ecology is the pig [145]. The disease may kill baby pigs but resistance increases with age and adult pigs commonly have an almost inapparent infection. There is, however, some nasal discharge and this is the source of infection for cattle. Pigs are often allowed to run with cattle in the mid-west to clear up the food which the cattle waste, and it is when this is done that trouble from pseudorabies may start. Some workers have recovered virus also from the urine of pigs and over quite long periods; others, however, have failed. Animals other than pigs are not naturally infectious. The ecology may be rather different in Hungary and elsewhere in Europe where older pigs are more likely to be infected and where the evidence that other animals always catch the disease from pigs is less convincing.

Brown rats may be affected, being readily infected by eating contaminated material, and it is likely that they may carry the virus from one farm to another. In Northern Ireland, where the disease is more commonly recorded than elsewhere in the British Isles, there have been cases in terriers which were used to kill rats on pig farms. Both rats and pigs, therefore, seem to be concerned as reservoirs of infection; whether either of them is the virus's original host we do not know.

Two other diseases will be mentioned in this chapter; they are of interest as illustrating the dangers of introducing domestic animals into regions where there are wild animals existing in peaceful equilibrium with their own viruses. There is a virulent form of swine fever in East, South and West Africa; it is now known to be caused by a different virus from that of the swine fever or hog cholera of Europe and North America. The African agent is probably a DNA-virus, while ordinary swine fever is probably related to myxoviruses. African swine fever is also known as wart-hog disease, as it was recognized that it turned up in domestic pigs where wart-hogs and bush-pigs were prevalent [48]. These animals are apt to visit farms in search of food at night. In them there is a chronic inapparent infection and the virus may be present in their blood over a period of several months. Just how the pigs pick up the virus from wart-hogs is uncertain; possibly it is by the mouth. A vector has been suggested but not proved. Once pigs are infected the virus spreads among them by contact. The environment becomes contaminated and premises once infected remain so for some time. A chronic viraemia is occasionally seen in domestic pigs. In 1957 the disease was somehow introduced into Portugal and Spain. Herds containing sixteen thousand pigs were affected and a vigorous stamping-out policy had to be initiated.

Lastly there is a fatal disease of cattle known as malignant catarrh. This, like pseudorabies, is caused by a virus related to Herpes. There is severe inflammation of mouth, nose and eyes and there may also be nervous symptoms. It occurs in Africa, Europe and North America, but it is not certain that the disease in more northerly countries is the same as that in Africa. Serious outbreaks occur in Africa; in Europe and America, on the other hand, the disease is sporadic; generally only one or two cases turn up in a herd in one season but they may appear on a particular farm in successive years. The infection in sheep is mild or inapparent and it is commonly believed that cattle are infected when in close contact with sheep. The real reservoir, in Africa, however, appears to be the gnu or blue wildebeest (*Gorgon taurinus*) [129]. These, like the wart-hogs infected with African swine fever, carry the virus in their blood for long periods: one in captivity had viraemia for eight months. Some wildebeest calves may be infected congenitally. Forty per cent of those one to two months old have viraemia and infection is transmitted to bovine calves by contact. Just how this happens is obscure.

Chapter 18

Evolution in Action: Myxomatosis

OUR KNOWLEDGE of the course of evolution has largely been obtained laboriously and indirectly, as by the study of fossils in successive geological strata. The virus of rabbit myxomatosis has given us an unexampled opportunity of observing evolution in action, an evolution which has involved not only the rabbit and the virus but perhaps other species also [58].

The virus is a member of the poxvirus family, closely resembling vaccinia virus, as the electron microscope reveals. Its terrible effects on wild rabbits are familiar to most people in western Europe and Australia. In order to understand its evolution we must consider it, together with some closely related viruses, as they exist in different continents, first America, then Australia and finally Europe. We must first bear in mind a fact or two about rabbits. They and their relations, the hares, were formerly classed as rodents but they differ in the arrangement of their teeth and in other ways and are now put into a separate order, the Lagomorpha. The Leporidae, the main family in the order, contains nine genera, of which only three concern us. There are a number of species of hares, *Lepus*, in Europe and North America, a number of *Sylvilagus* species in the Americas, and true rabbits, *Oryctolagus*, from Europe and North Africa. Tame rabbits are derived from this last, *O. cuniculus*, the only species in its genus. Unlike most of them, it mates and lives in underground burrows. It apparently did not reach Britain from southern Europe until the twelfth century; it was subsequently introduced into Australia, Chile and other parts of the world.

Infecting *Sylvilagus* species in America are three viruses, closely related to each other in immunological tests and giving considerable cross-protection against each other. The one we are chiefly concerned with is the myxoma virus of the tapetis (*S. brasiliensis*) of South America. Then there is the fibroma virus discovered by Dr R. E. Shope [144] and often called Shope's fibroma; it affects the eastern cottontail, *S. floridanus*, in the eastern United States. Finally, another related virus is endemic in the sage-brush rabbits, *S. bachmani*, of California [112]. All these viruses cause in their natural hosts localized swellings, fibromata, due to proliferation of connective tissue cells. The diseases are mild and non-fatal except experimentally in newborn animals.

The myxoma virus first came to notice when in 1896 Dr Sanarelli, an investigator in Montevideo, Uruguay, suffered the loss of most of his colony of tame rabbits (*Oryctolagus*) from an unknown disease. This was found to be due to a virus, the properties of which were worked out. The virus was readily transmitted to domestic rabbits but not to other species. Over ninety-nine per cent of infected rabbits died. After five to seven days' incubation there was a discharge from the eyes which swelled up until the rabbit could not see. Swellings appeared in other parts of the body, particularly on the head, round the base of the ears and round the anus and genitals; eleven to eighteen days after infection the rabbits died. The swellings were due to overgrowth of cells of connective-tissue origin, often having a stellate form, the whole mass soon undergoing gelatinous degeneration. The origin of the disease remained a mystery until, twenty years later, Dr Aragão in Brazil discovered that the infection could be transmitted by fleas and mosquitoes. After a further twenty years or so, Aragão recognized that the infection must have come from the native rabbits, the tapetis.

As already mentioned, the infection in tapetis is a relatively harmless one, the firm fibromata being a very different affair from the gelatinous swellings and associated fatal symptoms in domestic rabbits. The ecology among tapetis has not been fully worked out, but it seems likely that transmission in the wild is by mosquitoes, which contaminate their mouth-parts by biting through the swellings and then carrying virus mechanically to the next tapeti on which they feed.

The infection produced in North American cottontail rabbits by Shope's fibroma virus is in the naturally infected species very similar to that produced by myxoma virus in tapetis. An important difference is that on transmission to domestic rabbits the disease remains a mild and normally localized fibroma. In very young rabbits and in older rabbits treated with tar, X-rays, cortisone and in other ways, there may be generalized nodules and even occurrence of true tumours – sarcomas. The disease stands between the ordinary virus infections and malignant growths. It can be shown experimentally that the fibroma infection can be transferred by mosquito-bite. If the mosquito is the main transmitting agent, we have to face the same problem as with arboviruses as to how the virus survives in the winter when mosquitoes are inactive. The disease is, however, a chronic one and the fibroma may persist in an infective state for as long as ten months. All the same, the ecology is still obscure. Lesions are usually on the foot; they commonly appear in July and August: Dr Shope informs me that he has not yet succeeded in discovering a vector whose activities would satisfactorily fit all the facts.

The third virus of this family to be discovered is the one from California [112]. Cases of myxomatosis in tame rabbits had been turning up there since 1930; in 1960 the reservoir was run to earth in sage-brush rabbits (*Sylvilagus bachmani*) trapped near one of the outbreaks. In these rabbits the infection takes the form of a localized fibroma, as do the other two viruses affecting *Sylvilagus*. The effects of the Californian virus on tame rabbits differ from those caused by the other two; the disease in *Oryctolagus* is commonly fatal, but one does not see the swellings which myxoma virus produces. There seems to be a close, stable, association between the Californian virus and the *S. bachmani* and mosquitoes, particularly *Anopheles freeborni*. The virus was successfully transmitted by mosquito-bite to five different species of *Sylvilagus*, but a second such transfer in series by mosquitoes was only successful in the case of *S. bachmani*. In the other rabbits it was a dead-end infection. Similarly the Brazilian myxoma virus, though infecting *S. bachmani* did not produce enough virus to be carried on by mosquitoes: another dead end!

We may mention in passing yet another American virus of this family: this is one producing fibromas in grey squirrels. It is transmissible to rabbits but has not gone in series in any species except squirrels and woodchucks (*Marmota monax*) [95].

The story of how viruses cause fibromas in America serves merely as a background to the remarkable developments seen in Australia. It was Dr Aragão of Brazil who first suggested that the myxoma virus could be used to control rabbits where they had reached plague proportions, as in Australia. Well might the Australians welcome this idea, for rabbits, introduced on several occasions in the last century, had multiplied so successfully that they had made large areas worthless for rearing sheep and had ruined many sheep farmers, costing the country tens of millions of pounds. It was first established that the virus would be harmless for domestic animals other than rabbits and for native Australian marsupials. The early liberations of virus had no success, partly, no doubt, because the role of insect vectors was not appreciated; interest accordingly flagged.

New attempts were, however, made in 1950, infected rabbits being liberated at seven sites in south-eastern Australia. At first it seemed that these attempts, too, had failed; and indeed six of the seven did fail. But, when hope was almost lost, the seventh led to a flare-up in the disease in December 1950, near the site of the liberation and soon afterwards from a number of places in the Murray River basin, one of them nearly four hundred miles away. At last the importance of mosquitoes was realized, the spread being now attributed to the movements of *Culex annulirostris* along river beds. The disease continued to spread till June 1951, only to die down early in the Australian winter. But next summer the disease reappeared and extended over most of New South Wales and Victoria and even beyond. The work of Professor Fenner and his colleagues at the University of Canberra showed that a number of insects could carry the virus; but they only did so mechanically, acting as 'flying pins': there was no biological cycle as with arboviruses. *Anopheles annulipes* now seemed to be particularly important; this mosquito rests in rabbit burrows during daylight hours and, moreover,

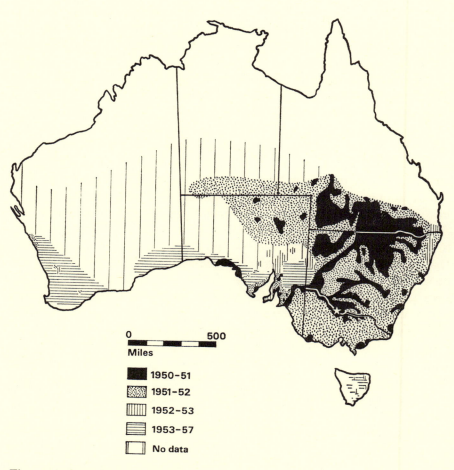

0   500
Miles

- 1950–51
- 1951–52
- 1952–53
- 1953–57
- No data

Figure 18. Spread of myxomatosis.

flies much more widely and is not so confined to the vicinity of water as is *C. annulirostris*. The virus has, however, spread better in years of heavier rainfall which have favoured various mosquito species. By 1952–3 a really wide spread was evident, millions of rabbits being killed, perhaps four-fifths of those in south-eastern Australia (figure 18). The increase in value of rural production was estimated at £50 million. Mortality among the rabbits at first was almost 99·5 per cent.

This, however, was too good to last and before long a number of liberations of virus showed poor results. The virus was killing fewer and fewer rabbits. From what has been written earlier in this book, such a result could have been predicted. No highly specific parasite can exterminate its host, for this involves its own suicide. Two things seem to have happened in Australia: the virus became less virulent and the rabbit became more resistant. The interaction of these two happenings gives us an illuminating glimpse into the evolution of infectious disease.

How the attenuated virus began to emerge first became clear from the studies of Mykytowycz [119]. A mosquito cannot pick up enough virus to infect another rabbit from the blood it sucks; it must bite through a myxomatous swelling. Virus wiped off thence on to its proboscis may remain there for as long as two hundred days. If a myxomatous rabbit, infected with a highly virulent virus, dies quickly, it will be available as a source of food for mosquitoes for a relatively short time. If, infected with a rather milder virus, it lives longer, it has a chance of contaminating more mosquitoes and that strain will consequently be spread more freely. If, however, the virus becomes too attenuated, the nodules will be smaller and there will be insufficient virus in them to be passed on to more rabbits. Thus, a virus of an intermediate degree of virulence will be favoured. Events soon showed that this was not killing rabbits fast enough to satisfy the farmers, so highly virulent strains were liberated. These did in fact, at first, succeed in killing more rabbits, but very soon the less virulent strains became the most prevalent ones and by the next season the situation was as before.

Professor Fenner and his colleagues have made painstaking studies of the changes in virulence taking place in the myxoma virus. It turned out that the percentage mortality produced in fully susceptible domestic rabbits bore a very close relation to the average time it took rabbits to die after infection. It was thus possible to form a good idea of the virulence of a strain by inoculating only five rabbits. Five grades of virulence were established according to mortality rate and survival time and figure 21, p. 158 shows how the virulence has shifted between 1950 and 1959 from grade 1 to grades of

Figure 19. Map showing sites of first appearance of myxomatosis in England. The first site (at Edenbridge) is marked 1, and so on.

154

greater attenuation [110, 111]. Most strains isolated since 1952–3 have been in grade 3. This kills off ninety per cent of rabbits, but rabbits are so fertile that their numbers soon build up again from the resistant survivors, especially in the winter months when mosquitoes and therefore the viruses are less in evidence.

This state of affairs greatly favours the second happening referred to – increase in rabbit resistance. Clearly an agent which kills over ninety-nine per cent of its victims puts a tremendous premium on possession of an in-born tendency to resist, and so it is not surprising that the genetic resist-ance of the rabbits soon began to increase; this did actually happen rather earlier than the Australians expected. Its existence was shown by experi-ments complementary to those which tested the virulence of the virus. In-stead of testing a number of viruses of various origins on the nearest ap-proach to a 'standard' rabbit, the Australians tested a standard strain of virus, one of intermediate virulence, on a number of wild-caught rabbits. Rabbits from one area were so resistant that only twenty-six per cent died when inoculated with a strain which killed eighty-eight per cent of 'ordin-ary' rabbits. The degree of resistance was greater in rabbits from areas which had experienced a larger number of outbreaks. On the whole the effect of changes in rabbit resistance was roughly equivalent to that due to changes in rabbit virulence. As already suggested, however, these two fac-tors can be expected to interact. If transmission is favoured by a particular degree of severity of infection, not too high, not too low, what happens if a nicely attenuated strain infects rabbits with high resistance? Clearly the virus will then produce too mild a disease for really free spreading and natural selection will favour a return to rather higher virulence; we shall then see also a return to the optimum disease-severity. Thus virus may vary, rabbits may vary, but natural selection will tend to produce a rela-tively stable disease-picture. Fenner points out that when we see the same sort of disease produced by any parasite year after year, we may tend to assume that parasite and host are remaining stable in their characters: actually both may be varying, only the disease remain-ing the same [58].

The net result of all this on Australian rabbits is that myxomatosis by itself is 'out' as a means of keeping the animals under control. It can only help the farmers there at best in a way supplementary to other measures. Fenner makes the interesting speculation that if only the virus had failed to overwinter, no attenuated viruses would have been carried over to the next year, there would have been no time in one season for rabbits to de-velop any appreciable genetic resistance; so the farmers by introducing virulent strains every year would have had a hope of keeping the rabbits down to relatively harmless numbers.

In New Zealand, introduced myxoma virus failed to establish itself, doubtless because of the absence of suitable vectors. On the other hand it spread successfully when let loose among the European rabbits which are a pest in Chile. Oddly enough it even got going in Tierra del Fuego, where neither fleas nor mosquitoes could be transmitting it.

We now come to Act III, in Europe. In June 1952, a retired French physician, hearing of the Australian successes, bethought him to introduce myxomatosis to control the rabbits on his estates near Paris. He certainly succeeded: not only did most of his rabbits die but the disease spread all over France and into Belgium, Holland, Germany, Switzerland, Italy, Spain and Great Britain. At first it was thought that the disease was spread by mechanical means, as for instance, on the tyres of cars which had run over and killed affected rabbits. But the spread to domestic rabbits soon carried conviction that an insect vector was concerned; and *Anopheles maculipennis atroparvus* seemed to be a mosquito largely responsible. The French authorities tried in vain to limit the spread of the disease, for rabbit-breeding is an important industry and the week-end *'petits chasseurs'* were much perturbed at the disappearance of the rabbits.

In the late summer of 1953 myxomatosis appeared on an estate in Kent and shortly afterwards at a number of other places in south-east England. How the infection reached Britain will probably never be known. After its arrival it was certainly spread by farmers wanting to get rid of their rabbits, and its appearance was then 'spotty' as might be expected. The fact that all the earlier outbreaks were in counties opposite the French coast suggests that it was not deliberately introduced (figure 19). There are two likely mechanisms. Rabbit fleas, though birds are never their permanent hosts, may be carried by migrating birds. Again, infected mosquitoes could well be wind-borne, as are many other insects, across the Channel. Attempts were made to contain the disease and it was fully expected that the disease would die down in the winter, as it had done in Australia, or even die out altogether. Quite the contrary happened: the disease smouldered along through the winter and slowly spread thereafter all over England and Scotland. By the end of 1954 nearly five hundred outbreaks had been reported (figure 20). In Australia myxomatosis had been regarded by farmers as a blessing: in France the *petits chasseurs* had thought it a curse: the sentimental English were largely swayed by the sight of rabbits suffering and dying from a very unpleasant disease. An order was made prohibiting the deliberate spreading of infection by moving infected rabbits.

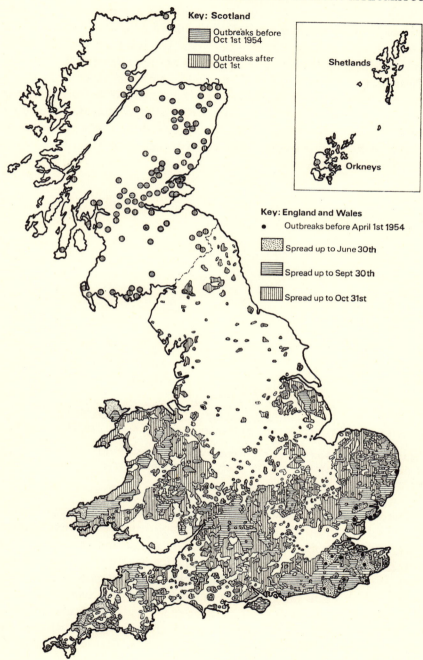

Figure 20. Map showing spread of myxomatosis in Britain during 1954.

Near the south coast *Anopheles atroparvus* was incriminated as a vector, spreading infection to tame as well as to wild rabbits, as in France. But in the country generally it soon became evident that it was not mosquitoes but the rabbit-flea, *Spilopsyllus cuniculi*, which was the main vector. This flea, which normally infests only rabbits, is of course active throughout the year, a fact explaining why myxomatosis is not a largely seasonal disease as it is in Australia. *Spilopsyllus* does not occur in that continent. Study of the flea's ecology, particularly its relations with its host, has been of consuming interest in its bearing on the differences between myxomatosis as seen in Australia and in Britain.

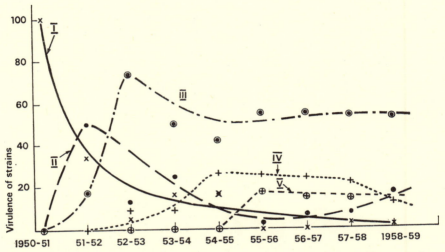

Figure 21. Chart showing the prevalence of myxomatosis virus of different grades of virulence in Australia over ten years. The one of intermediate virulence now predominates.

The disease was, as all know, dramatically effective in reducing the numbers of rabbits in Britain to a small fraction of their previous level, and in 1965, twelve years after the virus was introduced, the rabbits' numbers are still relatively low in most areas. This is in large measure because of intensive efforts to keep the rabbit population down by gassing and other means, a task made much easier when the rabbits are relatively few as a result of myxomatosis. At first it seemed that the virus's virulence was being maintained at a much higher level than in Australia: indeed it does still appear that attenuation has been more gradual and less complete in Britain. Samples of virus from different parts of Britain have been sent to Professor Fenner's laboratory in Australia for comparison with Australian viruses. Some results of the tests are shown in the following table [59].

Comparison of the virulence of naturally occurring strains of myxoma virus in Australia and Great Britain, at the time of the first epizootics and at various periods after this. Figures represent the percentages allocated to the virulence grades shown.

		VIRULENCE GRADE						
		I	II	IIIA	IIIB	IV	V	
		MEAN SURVIVAL TIME (DAYS)						
		13	14–16	17–22	23–28	29–50	—	
		CASE MORTALITY RATE (%)						
		99	95–99	90–95	70–90	50–70	50	
COUNTRY	YEAR							SAMPLE SIZE
Australia	1950–51	100	—	—	—	—	—	—
	1958–59*	0	24·6	29·2	26·1	14·0	6·1	130
	1963–64	0	0	26·4	42·6	25·6	5·4	129
Great Britain	1953	100	—	—	—	—	—	—
	1962	4·1	17·6	38·8	24·8	14·0	0·9	222

Grade III virulence, as it was shown in figure 21, has now been subdivided into IIIa and IIIb: the differences thus brought out seem important. It will be clear that nine years after the virus's introduction, viruses of grade I and II virulence are still turning up in Britain, while highly attenuated (grade V) strains are relatively rarer. Grade IIIa virulence with a 90–95 per cent kill, seems to preponderate in Britain compared with grade IIIb (70–90 per cent kill) in Australia. A possible reason for the retention of higher virulence has been put forward [12]. Fleas are less mobile than mosquitoes. Should they tend to remain on a sick rabbit, but quit a dead one, there would be earlier movement of fleas and more likelihood of rapid spread from a rabbit infected with a quickly lethal virus than from one living longer. There would then be a factor favouring the distribution of more virulent viruses, in contrast to the opposite tendency in Australia. This argument has less force than was at first thought. It was found possible to mark individual fleas by clipping off the end of one or other of the six legs and to follow the distribution of the marked fleas on rabbits in an enclosure. Considerable interchange of fleas among healthy rabbits was found to occur sometimes within as short a time as three days [113]. Evidence is at present lacking as to whether British rabbits are acquiring any genetic resistance to the virus, but the matter is being investigated.

Another puzzle has presented itself. Outbreaks of myxomatosis have been appearing here and there all over the country during the past decade. Rabbits have virtually disappeared and then come back again, only to be decimated some time later by more myxomatosis. Where has this virus come from? Sometimes, no doubt, as a result of deliberate illicit introduction; but this has not, it is thought, been a general explanation. It has been shown that some fleas which have fed on an infected rabbit can survive underground for as long as 105 days and still transmit myxomatosis. A certain amount of persistence of infection can be accounted for in this way [39].

Ticks survive longer in nature than do fleas and might be expected to be more efficient in maintaining infectivity over longer periods. They are not very numerous on British rabbits, though in North Africa they infest them freely and may well be important vectors. However, even a few virus-bearing ticks might be responsible for introducing virus into British rabbits which are recolonizing an area from which they have temporarily disappeared. The entomologist Miriam Rothschild [137] has shown recrudescences of the infection following movements of sheep into a virus-free area: sheep are commonly infested with ticks. Other introductions may have been from birds, particularly carrion eaters which may have acquired fleas when feeding on dead rabbits. Virus may also be spread by fleas on hares; these animals cover much longer distances than do rabbits.

This brings us to another series of phenomena. The ecology of the myxomatosis virus involves also the ecology of many species of animals and plants. Obviously it affects primarily the rabbits. Not only does it drastically reduce their numbers but it seems to have led to changes in the habits of survivors. Rabbits in Britain no longer live mainly in deep burrows in warrens; they are to a greater extent than before surface-living animals, as are American cottontails. No doubt this is partly due to the falling-in of ancestral burrows when rabbits were absent. Possibly also the acquisition of these new habits would be unfavourable to the rabbit-fleas which breed in rabbit nests and burrows. The disappearance of rabbits led to an increase in the population of hares, which benefited from removal of a competition and became, more than previously, frequenters of woods as well as open fields. Of course myxomatosis hit the rabbit-fleas as hard as it hit the rabbits, for their source of livelihood was taken away. The breeding of rabbit-fleas is intimately linked to that of their hosts. The ovaries of female fleas only develop when they feed on pregnant does or very young rabbits; they require stimulation by ingested hormones from their hosts. Maturation does, however, tend to occur when the fleas feed on myxomatous rabbits. There have been indications that rabbit-fleas are now less rigid as regards their diet, for they have in places been found more frequently than ever before on hares, though there is no evidence as yet that breeding in relation to this host is occurring [137]. A few instances of myxoma infection have been recorded both in *Lepus europaeus*, the brown hare, and in *L. timidus*, the blue or mountain hare of the north. It seems that while most hares are resistant, just a few are susceptible, in contrast to rabbits where there are a few resistants among hundreds which are liable to infection.

Other vertebrates have been affected, too. Man has gained from the benefit to agriculture. Buzzards, on the other hand, which fed largely on young rabbits, were reported as doing very badly after myxomatosis's arrival. Foxes, contrary to expectation, did not suffer, being adaptable and relatively omnivorous animals. Apparently they ate more voles, so these were losers from the change.

The effects on vegetation have been tremendous [166] (plate 37). Some orchid species, normally chewed down by rabbits, have appeared in places where they were not known to occur. The consequences of the disease were, however, by no means all for the best: scrub, released from control by rabbits, has been able to cover large areas of open downland. Doubtless, too, vegetational changes have affected populations of birds and of insects.

As a footnote, a recent discovery must be mentioned. The myxoma family of viruses was until lately thought to be a product only of the New World. Then in 1961 some Italian workers reported on a nodular disease of hares in the Po Valley. The virus they isolated proved to be immunologically related to the myxoma fibroma viruses of America. Further it appeared probable that the disease was identical or similar to a 'sarcoma of hares' reported from the Rhineland in 1909 and since almost forgotten. No doubt the study of the natural history of myxomatosis has many remarkable things yet in store for us.

Chapter 19

Herpes and more about Latency

LATENT INFECTION with a virus was referred to in chapter 1 as being a common state of affairs, and we have had to refer to the matter since as something of great importance. The forces of natural selection have generally acted to favour less virulent viruses; a way in which this apparently came about in the case of myxomatosis was discussed in the last chapter. Doubtless somewhat similar mechanisms have operated in the case of other viruses. Ability to spread actively and aggressively favours a virus when it has a chance to extend its empire; in contrast, its long-term survival is favoured when it is modest and unassuming. So it happens that some viruses, including the first two to be considered in this chapter, seem to thrive in a system where activity and quiescence alternate. Such an alternation was apparent also when we were considering colds and influenza in chapters 5 and 6.

A mechanism which probably comes into play concerns interferon (see p. 36). It has been observed in growing several kinds of viruses in tissue-culture that the cells in the culture are nearly all killed, but that some survive and grow out freely. In a culture of such surviving cells virus may persist for weeks or months. In some instances it seems that a few cells become infected and die, liberating virus but also producing interferon which protects the rest of the cells or nearly all of them; when such cultures are treated with antiserum the culture is cured of its infection, for the virus is neutralized as it comes out of a cell and before it can infect another one. In other instances it seems that most of the cells are infected with virus but are not damaged: in such cases antiserum fails to effect a cure. What seems to happen is that interferon is produced in quantities sufficient to protect the cells from harm but insufficient to stop virus growth altogether: at the same time enough virus is produced to keep the strain going but not enough to damage the cell. The same sort of thing may happen in the living animal.

It is known that different viruses and even strains of one virus differ in their ability to evoke interferon-production and in their resistance to its restraining powers; in general the less virulent strains are more sensitive. It is possible, though this is only hypothesis, that sensitivity or otherwise to interferon is a fairly labile character; environmental influences might switch the virus between a more sensitive and a more resistant phase, thus enabling it to attack and spread more actively at times and to remain quiescent when that suited it better.

Let us now consider a few virus infections in which latency is an important feature. The classical example is the virus of herpes simplex which causes 'fever blisters' in man. This eruption is apt to come out in some people usually at the corner of the lips or on the nostrils; others are never troubled. Various stimuli activate the infection in those who are susceptible. It may be a common cold or a fever, or the eruption may be brought on by eating cheese, taking certain drugs or by menstruation. Curiously enough it is much commoner in some fevers than in others. In lobar pneumonia, malaria and cerebro-spinal fever it has been reported in forty per cent of cases. In other infections it is much rarer, particularly so in typhoid fever and smallpox. Though commonest around the lips and nose it may occur in other places. When the eye is affected, it is a more serious affair as it may lead to opacities of the cornea. One girl had an eruption on the buttocks at about the same time every year.

There is now no doubt that 'habitual herpetikers', as they have been called, have a latent infection, usually silent but occasionally active. There is some doubt as to where the virus is lying hid in between whiles; it could be in the cells in the skin or in the local nerves or nerve ganglia. The last suggestion derives from knowledge that herpes virus has great affinity for the nervous tissues of rabbits and other experimental animals. Two workers in France failed to isolate virus in between attacks from the skin of the favoured site of the eruption. In another experiment skin-grafts were taken from such a site and the patient was given a drug known to be likely to provoke an attack. The eruption duly broke out – at the usual place and not on the skin grafted on to a new site. Virus has, however, on numerous occasions been recovered in small quantities from the saliva of patients during periods of virus-quiescence.

Two features of interest may be mentioned. 'Habitual herpetikers' have quite good antibodies to the virus in their sera; those who are naturally immune have none. This is contrary to the usual rule where immunity goes hand-in-hand with presence of antibody. Then again, susceptibility and the associated antibody are found to be much commoner in lower socio-economic groups than in those better off. Dr E. A. Carmichael and I [10] found this to be so when comparing out-patients in the City of London with medical students, the latter being, if not always better off, at least from more affluent families! Others have since confirmed the finding.

163

The underlying facts have now become clear [49, 35]. Infection usually takes place early in childhood: if one escapes then one develops a resistance which is not due to antibody-formation and one is likely to escape altogether. The resistance is not absolute, for transfer of infection may occur between engaged or married couples: big doses of virus probably break the resistance down! It is also possible to infect the skin of human beings experimentally; this seems to be a good deal easier to achieve in the case of the herpetikers than with the naturally resistant people; it is also said to be easier in the former at a time of natural recurrence than in between attacks. The initial infection in children is usually a silent one, but may take the form of stomatitis, an infection involving all the mouth; with this there is commonly fever [49, 35]. When the infection subsides, virus apparently manages to persist in the odd corner, waiting to be woken up by the cold, cheese or other stimulus. A curious fact is that young children rarely get their first attack before the age of one. This is doubtless accounted for partly by maternal immunity, that is antibody passed from an immune mother to her child by way of the placenta, but it is surprising to find immunity of such an origin persisting for as long as a year. Most studies of the subject have been in populations where incidence of infection and of antibodies in mothers was very high; comparable studies are not reported where most mothers were naturally resistant in the absence of antibody. It seems fair to conclude that periodic activation of a latent infection helps the spread of the virus: certainly more infection must occur from the large amounts of virus shed from herpes blisters rather than from the small quantity to be found at times in saliva between recurrences.

The herpes simplex virus can do other things to man than producing fever blisters and stomatitis. It may cause a generalized fatal infection, and this may occur in very young children. Such serious illness sometimes accompanies an extensive skin infection with the virus in children with widespread eczema: presumably the virus readily attacks the eczematous lesions. A number of fatal cases have been reported in rather older (nine to sixteen months old) children in South Africa [17]. Almost all the infants were suffering from malnutrition. The damage in such cases is especially to the liver and adrenal glands.

It is hardly surprising that herpes virus can also damage the nervous system of man, for, as already mentioned, the virus readily attacks the nervous system of rabbits, mice and other species infected experimentally. There is evidence that the virus travels from a peripheral site of inoculation partly by the blood-stream but also partly along nerves. It was once thought actually to go along the axons themselves, the prolongations of the nerve-cell; it now seems likelier that it is spreading by infecting certain interstitial cells in the nerves. Anyway the result in the animals is commonly fatal encephalitis. Herpetic infection of the nervous system in man may take the form of a meningitis or an encephalitis or both: it is a rare complication of mouth or nose lesions of the usual type and it often, but not always, occurs in young children.

Light still needs to be thrown on one aspect of the infection – why many people escape permanently if not infected in early childhood. This is, of course, only one example of a common phenomenon: many animals acquire as they grow older a resistance to infection with viruses of various kinds, and this quite apart from development of the specific immunity which is associated with antibody in the blood. Some recent work affords a clue. Herpes is one of the viruses which can spread from cell to cell in cultures, and probably also in the living animal: sometimes it does it by causing adjacent cells to fuse together. R. T. Johnson, working in Australia [90], was able to infect macrophages – large scavenging cells with single nuclei. This he could do when they came from people of all ages, but only when they came from young people would the virus spread to other cells in contact with those first infected.

The virus of chickenpox or varicella is not related to the other poxviruses but to the herpesvirus family. Like herpes simplex it seems to have evolved an ingenious mechanism for survival. Chickenpox behaves in general like the other infectious diseases of childhood. It is rather less infectious than measles, has an incubation period of fourteen to sixteen days, rarely longer and is usually mild, without complications. Its importance rests in part on the fact that it is sometimes difficult to distinguish from smallpox, and if smallpox turns up and is dismissed as being merely chickenpox, the consequences may be serious. At one stage of its development the rash of varicella consists of little vesicles or blisters. Another, apparently very different disease, shingles or herpes zoster, also shows itself as a rash with blisters. But these blisters are very localized, occurring only over the area of distribution of a particular nerve or group of nerves. The area is that supplied from one of the ganglia in the spinal cord. Appearance of the rash is preceded by very considerable pain, due to the inflammation of the ganglion concerned. The pain may lead to the erroneous diagnosis of pleurisy or appendicitis, but the smart doctor can gain much credit if he spots what it really is and prophesies that a shingles rash will shortly appear over the painful area.

It was noted, back in 1888, that cases of chickenpox might turn up in children in contact with a person having shingles. There was at first a lot of scepticism about a possible relationship. Chickenpox, it was pointed out, is a disease of children; shingles attacks mainly adults. Chickenpox comes in epidemic waves; cases of shingles are fairly uniformly scattered throughout the year. It was, however, finally settled that the viruses of varicella and shingles are indeed one and the same: vesicle fluid from shingles inoculated into children who had not previously had chickenpox, came down with typical varicella and passed it on to other children. When it was learnt how to grow varicella virus in tissue-culture, laboratory tests quickly confirmed its identity with that of herpes zoster. It seems that chickenpox rarely gives rise by cross-infection to cases of zoster, nor does zoster breed zoster: it is a one-way traffic – zoster to chickenpox.

There is now general agreement as to the explanation. After an attack of varicella, the virus goes to ground in the nervous system, apparently lying doggo in posterior root ganglia of the spinal cord. Later, usually many years later, some stimulus activates it, the stirred-up virus causes inflammation in a ganglion, usually only in one, spreads centrifugally along the nerves to the skin, where it produces its local vesicular rash. The stimuli are less easily identified than are those activating herpes simplex. Poisoning by arsenic or lead, syphilis and cancer are influences referred to in the textbooks; recently leukaemia and irradiation have been prominent in this connection. Most often, however, no precipitating stimulus can be identified.

Dr Hope-Simpson, a general practitioner in Gloucestershire, has recently reviewed 192 cases seen in his practice in the course of sixteen years and has thrown new light on the subject [80]. He fully concurs with the view that zoster represents a reactivation of dormant varicella virus. Two points of special interest emerge, concerning place and time respectively. Zoster rashes occur most commonly on the chest or abdomen or on the face, less frequently on the back of the head, neck or limbs. The former are just the areas favoured by chickenpox rashes: their centripetal distribution is one of the features which help to distinguish varicella from the centrifugal rash of smallpox.

As to time, zoster attacks rarely in children, more and more commonly as he gets older till octogenarians got an attack-rate which in Hope-Simpson's figures was fourteen times as high as that in children. It is suggested that you are not a candidate for zoster until you have already had chickenpox, and even then not until the antibodies generated by the attack have had a chance to wane somewhat. They may be stimulated by chance contact with varicella virus or by abortive attempts by the virus to break out of its prison in a ganglion. Finally they fail to restrain it and zoster results. Older people have fewer contacts with the young, who are most likely to shed small immunizing doses of virus in their direction; so their immunity is especially liable to fade. It is reasonable to suppose that this whole mechanism favours the varicella virus, for it ensures that a fresh source of infectious virus will be provided every so often, in case the child-to-child chain of varicella-infection peters out. Hope-Simpson raises the question of why chickenpox should have evolved this device while measles and other infectious fevers have not done so. He suggests that varicella is an age-old, specifically human parasite; it does not infect other animal species and has no very close relation which does so. When prehistoric man was living in family groups of at most thirty to sixty persons he probably had infrequent contact with other groups, so varicella when it got going would use up all susceptibles in a few weeks. It has been estimated that a population of about 200,000 would be necessary to keep such a virus going indefinitely. So a mechanism for perpetuating the virus, such as varicella seems to have evolved, would have to be developed if the virus in a small community were not to become extinct. It is quite likely that smallpox and measles did not become parasites of man until later on, when man had begun to live in larger communities. Smallpox may well have been derived from one of the closely related poxviruses affecting his domestic animals; measles may have come from one of its near relations, dog-distemper or cattle-plague (rinderpest). Man's social evolution from a cave-dweller to a townsman is intimately tied up with the acquisition and successful development of his viruses and other parasites.

A particularly knotty problem is presented by viral hepatitis in man. In this infection damage to the liver commonly leads to jaundice, though there are also many mild cases with gastro-intestinal upsets but without actual jaundice. Two conditions are concerned, infective hepatitis due to a virus called A and serum hepatitis caused by virus B. A number of claims have been made to have grown a virus, especially virus A, in tissue-cultures, but none of them has been established as certainly genuine. Workers in other laboratories seem to be unable to confirm the results described by the first describer of a cultivated virus; the viruses isolated and grown by different workers mostly have different properties, often not the same as those of the naturally occurring virus. Thus the hepatitis viruses are amongst the very few which have neither been transmitted to experimental animals nor definitely grown in tissue-culture. Our knowledge of the virus's properties thus depends on experiments with human volunteers. Such tests cannot be lightly undertaken, as they could be with the common cold. Hepatitis can be a very unpleasant disease and a few fatalities among volunteers have occurred. Knowledge about the viruses has accordingly been hard to come by.

The story of infective hepatitis (virus A) is fairly straightforward. The virus infects mainly children, especially in autumn and early winter; it is present in the blood and excreted in the faeces; transmission is accordingly by the intestinal-oral route. The disease is particularly common in the Middle East and around the Mediterranean; here and elsewhere it has caused much illness among troops coming from other countries. Outbreaks have occurred as a result of eating contaminated oysters or clams. Milk- and water-borne epidemics are also on record. The virus is apparently a fairly small one and it withstands heating to 60°C better than most viruses do. The incubation period after infection is usually between fifteen and forty days, sometimes as long as fifty days. It comes into this chapter on latency because virus has been recovered from the blood as long as sixteen days before the appearance of jaundice, and because it occasionally persists in the faeces in chronic cases for many months. But infectious hepatitis has to be considered here even more because of its puzzling similarities to and differences from serum hepatitis, caused by virus B.

The symptoms of this second kind of hepatitis are precisely the same as those of infectious hepatitis; moreover the physical and chemical properties of virus B, so far as they have been studied, are the same as those of A. The chief, most important, difference is that B is transmitted only by injection of human blood or plasma or by introduction of human material into the skin or deeper tissues in some other way. Infection does not occur through inhaling or swallowing the virus, and the virus is only to be found in the blood, not in the faeces. Another difference from infectious hepatitis is that the incubation period is longer, 50 to 160 days instead of fifteen to forty. Other differences are that incidence is chiefly in adults and is equally great at all times of the year.

The virus seems to be present in the blood of a certain number of normal people—0·78 per cent according to one estimate; and it may be continuously or intermittently present over a period of years [97]. A few of these virus-carriers ultimately come down with jaundice, but this seems to be exceptional. Nowadays an enormous amount of blood is used for transfusions, and if blood from many persons is pooled, there is obviously quite a chance of including some from a carrier and so introducing virus B into people requiring transfusion; naturally this does them no good at all. As little as 0·01 c.c. of virus-containing blood can infect. The risk can be lessened by not pooling bloods more than is necessary and by keeping careful records of donors so that carriers can be identified and not allowed to give blood again.

Unfortunately we know of no safe way of ridding blood of virus B without spoiling it. Stable plasma protein, prepared from blood, is however safe and is as good as whole blood for some purposes. During the late 1930s and early 1940s thousands of soldiers and others developed serum hepatitis after being given yellow-fever vaccine, which at that time contained human serum as a constituent. Naturally the serum was eliminated as soon as the source of the trouble came to light. Another important cause of infection is the use of syringes for inoculating more than one person without adequate sterilizing in between. Although there have been numerous publications drawing attention to the danger, many doctors and others, I regret to say, still 'sterilize' their syringes and other instruments by immersing them in spirit. This is quite useless for getting rid of virus B, which only succumbs to autoclaving or prolonged boiling. Since virus A may, as mentioned earlier, be present in the blood of apparently normal people, it, too, may be transferred with syringes, as virus B is. This fact helps to confuse matters even more than before!

Infectious hepatitis is followed by considerable immunity to the virus, but several attacks do occur; possibly there are different serological types. Virus A infection does not lead to immunity against virus B. The latter infection also leads to some immunity to the same virus; immunity may be less than in the case of A, as is seen by the fact that older persons are as liable to be infected as younger ones. Infectious hepatitis, as the name implies, spreads by close personal contact as well as through the medium of water and other agencies. It is doubtful whether serum hepatitis is infectious at all; it would be hard to see how it could spread seeing that it is only present in blood, not in faeces or other excreta or secretions. True, a biting insect could do the trick, but there has never been the slightest suggestion that insects were concerned.

This brings us to the 64,000-dollar question: what is the natural method of transmission of virus B? No evidence exists that it is congenital, insect-borne, passed by respiratory or faecal routes or in any other of the ways we have yet encountered! The only vector seems to be man and his syringes, and one can hardly imagine a virus so evolving as to fit into a system involving only those. A possible solution has been suggested. Viruses A and B are alike in most respects and certainly produce similar symptoms when they do attack. Perhaps after an apparent, or an inapparent, infection with virus A, the surviving virus may pass into a phase (B) in which it is no longer reactive with anti-A antibody and no longer excreted into the intestinal canal, rarely being activated to do harm and not readily eliminated. Such an explanation might fit the facts, though goodness knows what use it would be to virus A to get itself into such a non-infectious state.

A virus-infection with some analogies to viral hepatitis of man is equine anaemia or swamp fever of horses. This disease may be acute or chronic, in the latter case with a series of acute attacks of illness with remissions between. There may be virus in the horse's blood over a period of years in between the attacks. There is on record a case of chronic inapparent infection in a man whose blood remained infectious for horses over several years [126]. The horse disease is commoner in swampy districts; *Stomoxys* (stable-flies) and *Tabanids* (horse-flies) have been suspected of carrying the disease mechanically on their mouth-parts and transferring it with their bite.

Horses also are reported to have developed a highly fatal hepatitis after injection with serum or tissues from other, apparently normal, horses. No infectious agent was recovered, but this may have been a conditon comparable to the serum hepatitis of man.

The last viruses to be considered in this chapter are three or four causing very slowly progressing diseases among sheep in Iceland [149]. In 1933 twenty sheep, mostly rams, of the Karakul breed were imported into Iceland from Germany. They remained free from illness during a quarantine period and were then distributed to a number of farms all over the island. But next year or the year after all sorts of troubles began, starting in the places where the German sheep had been introduced.

There was, first, a slowly progressing pneumonia called 'maedi'; this, in the Icelandic tongue, means dyspnoea or difficult breathing. Besides progressive difficulty in breathing the affected sheep showed loss of condition and anaemia and they ultimately died or had to be destroyed. The incubation period before symptoms appeared was one to three years, but it is evident that the infection really began much earlier. The white cells in the blood increased in numbers and lambs killed as soon as one month after infection showed early changes in their lungs. Only when these had become quite extensive did sheep show symptoms. These were especially apt to begin after periods of bad weather or other stresses. Between 1939 and 1952 at least 150,000 animals were lost from maedi; many farms were losing ten to twenty per cent of their sheep every year. Something drastic had to be done and the only available remedy was a policy of slaughter. So between 1944 and 1952 all sheep in the affected areas—about 300,000—were destroyed, and 180,000 healthy sheep brought in. This seemed to end the trouble, but in 1954, alas, some maedi began to turn up again in the same districts. A strict slaughter policy has so far prevented it from getting out of hand.

Besides maedi, the German sheep seem to have introduced another slow lung infection called pulmonary adenomatosis or 'jaagsiekte'. This last name comes from South Africa: it means driving-sickness and is so-called because the sheep get short of breath when driven. Symptoms are much the same as those of maedi but the changes in the lungs are very different: they consist of nodules of outgrowing epithelial cells, which increase until they involve a great deal of the lungs. The disease occurs sporadically in many parts of the world. It was apparently eliminated from Iceland as a result of the slaughter policy for maedi.

There has also turned up a slow disease of sheep involving the central nervous system, usually going on to fatal paralysis. This was called 'rida'; early descriptions are confusing because two nervous diseases were involved and had not been separated. The true rida is almost certainly the same as the scrapie known from Britain and especially from Scotland. It derives its English name from the fact that in some forms of the disease the sheep suffer from severe irritation and rub against gates and fences, often removing some of their wool in the process (plate 38). Other strains tend rather to cause sleepiness. The virus is a remarkable one in that it withstands boiling and differs in other properties from the generality of viruses. The natural incubation period is from one to four years. The infection may be latent and was accidentally transmitted to many sheep in Scotland because it was present in a vaccine against another virus disease, louping-ill: this had been prepared from sheep brains containing unsuspected scrapie virus. It is believed that scrapie may be introduced into flocks by latently-infected 'scrapie-rams'. Sheep which were apparently normal introduced scrapie into Canada and thence into the USA, also into Australia and New Zealand, just as infected 'normal' sheep brought new viruses into Iceland. There is no evidence however that the German rams were responsible for bringing scrapie into Iceland.

The other paralytic Icelandic disease, 'visna', is, however, likely to have come in with those disastrous German rams, as it turned up in the same areas as the maedi. It has an apparent incubation period, like maedi, of two years or so, but, as with maedi, the disease process really begins much earlier. Increased numbers of white blood cells are found in the cerebrospinal fluid as early as one to three months in sheep which have not yet developed any symptoms. Visna virus has now been grown in tissue-cultures, and a method was developed of measuring antibodies against the virus in these cultures; it was then discovered that antibodies were present not only in those of visna sheep but also in those with maedi. Other evidence has since come along indicating that these are pulmonary and nervous forms of one disease or are caused by variant forms of one virus infection [168]. However, visna has not reappeared since the slaughter policy against maedi was instituted, even though the latter has turned up again.

An important point which appears from all this work is that maedi and visna almost certainly came from those German sheep, yet neither disease is known in Germany. So presumably the infections there were wholly latent. Why, then, did the diseases appear in Iceland? Two suggestions may be made. Owing to harsh winter conditions Icelandic sheep have to be kept within buildings there in a way which is not necessary in warmer climes. Then, again, sheep in Iceland had had no contact with sheep from other countries for many centuries. An infection to which sheep elsewhere had built up a resistance might have serious effects on a population with no experience of it. One recalls that measles was a terrifying and often fatal disease when introduced into the Faeroes and into Fiji, islands whose inhabitants had not been exposed to the virus for long periods of time. Much the same is seen with some bacterial infections, especially tuberculosis.

Chapter 20

Congenital Virus Infections

CONGENITAL VIRUS INFECTIONS are those in which the virus is present in the tissues at birth: they may cause disease or be inapparent. Ludwik Gross of New York coined the term 'vertical transmission' to describe those cases where the infecting agent is transmitted not to neighbours but from parent to offspring [68]. He was dealing particularly with a leukaemia of mice; we shall come back to that along with other examples in the cancer field in chapter 22. Mention has been made previously (chapter 12) of the transmission of some tick-borne viruses via the ovum to a new generation of ticks.

At first sight it might appear that vertical transmission was something ideal from the virus's point of view: the line can be carried on with no risks from the hostile outside world. Well, it probably is a good idea, but only if supplementary, as in the tick-borne infections, to another method of transmission. In a stable population a pair of animals or birds only leaves behind, on the average, two surviving descendants; so viruses transmitted only in a vertical manner cannot expect to extend their empire—their dominion remains the same and may well shrink when things do not go well.

The first example to be considered in this chapter is a disease of man which is very common but of which few laymen have ever heard; it is called cytomegalic inclusion disease or salivary virus infection (cytomegalic means 'large-celled') [181]. There is a whole family of viruses with similar properties and habits, members of the larger herpesvirus family [156]. They are very specific for particular hosts, that is, they are not as a rule transmissible from one host species to another. Besides the human virus there are others infecting guinea pigs and mice—and those are the ones which have been most studied: yet others have been found in rats, hamsters, field-mice, moles, monkeys, chimpanzees, dogs, fowls, sheep, pigs and opossums.

They are particularly apt to settle in the salivary glands, where they grow in epithelial cells, often of the salivary ducts, and also in the kidneys. The cells affected are commonly enlarged to several times their normal size and they contain large inclusion-bodies in their nuclei, similar to those induced by other members of the herpesvirus family (see photograph on cover). In these situations they may persist for very long periods and virus may be excreted in the urine or liberated into the saliva.

The infection in man is usually a completely silent one; it is probably common, since eighty-one per cent of people over thirty-five have been found to have developed antibodies to the virus. The typical inclusion-bodies have been found in the salivary glands of small children dying from various causes; figures of eight to thirty-two per cent have been quoted for their incidence.

Rowe and his colleagues [138] found antibodies to the virus in nearly half of forty-seven children in a residential nursery in the USA, and of those who were positive the majority yielded virus either from mouth-swabs or from urine or both. Virus could be repeatedly recovered from six of the children over a period of several months. The infection may thus be silent and chronic. The story may, however, be a very different one. The virus may be passed to the children before birth, and not infrequently generalized disease follows such congenital infections. There is then very often jaundice which is present at birth with enlargement of liver and spleen, and encephalitis also occurs; such infections are usually fatal. Similar symptoms may develop in babies two to four months old; these may recover but often with severe physical and mental retardation. Here also we are probably dealing with congenital infections, but ones developing later than usual. Silent infections may be activated later in life: thus typical inclusions of cytomegalic disease have been found in the lungs in fatal cases of pneumonia after whooping-cough. The virus has also been found in the urine in cases of leukaemia or other malignant disease especially when occurring in childhood. Activation of viruses in the course of cancers will have to be considered again later (see p. 181).

A particularly interesting example of congenital infection is afforded by lymphocytic choriomeningitis, commonly abbreviated to LCM. The house mouse is the natural host of this virus; it has also been recovered from man, monkeys, dogs and guinea pigs, but these may have picked up the virus from mice. People probably contract the infection by inhaling dust contaminated with the excretions of mice; several cases have occurred in a single badly mouse-infected household. The disease in man may take the form of a simple influenza-like fever or there may be a mild, severe, or even fatal meningitis.

It is, however, with the disease in mice that our interest lies. Knowledge of its epidemiology is based largely on the studies of a German veterinary worker, Erich Traub, working at the time in America [171]. Traub studied a colony of mice infected with the LCM virus. Transmission from mouse to mouse took place in one or other of two ways, by contact or congenitally. Some pregnant mice were carrying the virus in their blood and these infected the young in the uterus; it was not known whether the actual ova were infected or whether infection came from virus in the uterus or placenta. The baby mice frequently became ill and some died within a month of birth; most, however, had only an inapparent infection. Mice born from uninfected mothers picked up the infection soon after birth and these had milder infections, manifested only by a rather slower growth-rate. This was the situation in 1935. In 1937, two years later, a change was clearly evident. All mice had become chronic carriers with virus constantly present in the blood in considerable amounts and all baby mice were infected *in utero*. However, instead of showing considerable mortality, they now all had subclinical infections; it is possible but not certain that they grew and bred rather less well than uninfected mice. Traub wrote of the virus 'one might call its present relationship to the host a "perfect parasitism"'. The result had apparently come about as a result of at least two changes. The virus had become less harmful to embryonic tissues so that the young were born without being seriously harmed; it had also become more contagious, a fact explained by the presence of more virus in nasal secretions in the congenitally infected mice. Possibly also the mice had become genetically more resistant.

It is evident from all this that mere presence of a lot of LCM virus in a mouse's blood and tissues does not make it ill. It is rather the reaction of the body against the virus which leads to local damage and to symptoms. These 1937 mice did not develop antibodies and acquire resistance of the conventional type. The reason, first suggested by Sir Macfarlane Burnet, involves a phenomenon we have not yet had occasion to mention, that of 'immunological tolerance'. The body, as we all know, reacts against foreign proteins, particularly by making antibodies against them. Now, our bodies contain very many different proteins in the different tissues; why, then, do we not make antibodies against our own proteins? A clue came from observations on twin calves. These may have before birth a common circulation via the placenta, the same blood circulating round both calves. Sometimes the calves are of different blood groups, and then contrary to the general rule, they do not make antibodies against the 'foreign' blood of their fellow-twin; their immunological system has a 'tolerance' for such proteins. It seems that animals do not make antibodies against any proteins encountered very early in their development and thus they have a mechanism for distinguishing between 'self' and 'non-self'. This leads to some queer results: if really foreign proteins such as an LCM virus reach the developing baby mice early enough, the defence mechanism is deceived into regarding it as 'self' and so no antibodies are made against it then or ever after. The damage it inflicts on the mouse's cells is trivial and so mouse and virus continue in peaceful coexistence. It may be mentioned in passing that this phenomenon of immunological tolerance lies at the heart of the problem of successful transplantation of kidneys or other organs or tissues. It is likely to be involved, also, in some other congenitally acquired virus infections in animals.

Perhaps the most striking examples of congenital virus infections are associated with viruses pathogenic for insects [154]. Most of those known infect larvae of butterflies and moths. Many of them, as mentioned earlier (see p. 123), form large protein inclusion-bodies (polyhedra) within which virus particles are embedded. In some infections the inclusions are found in cell-nuclei, in others in cytoplasm; this is a difference which proves useful in interpreting some experimental results. There was argument for a long time as to whether these viruses were very host-specific or were transmissible to other species of insect. It was asserted that instances of apparent cross-transmission were due to activation of latent viruses. It is now established that both things can happen; some viruses really cross to other species; others stir up dormant viruses in the inoculated larvae. Smith and his colleagues, working in Cambridge, fed larvae of privet hawk moths (*Sphinx ligustri*) with virus obtained from a polyhedrosis from one or other of six other moths. In some instances the inoculated nuclear disease was transmitted, killing the foreign larvae by the eighth day; but virus from one source, gipsy moth larvae, took longer to produce results and when the hawkmoth larvae did die it was from a different virus producing inclusions in the cytoplasm. Apparently the gipsy moth nuclear virus was not good enough to get going by itself, but did stir up a latent infection with a cytoplasmic virus. Among other similar instances a nuclear virus from painted lady butterflies stirred up a fatal infection in winter moth larvae; this again was due to a cytoplasmic virus [155]. On the other hand cytoplasmic viruses have not been recorded as activating nuclear ones.

There is no dispute that these latent viruses are commonly transmitted via the egg from one generation to the next. Indeed, geneticists and others rearing lepidoptera in captivity rather expect that after two or more generations their stocks of larvae will be decimated by viruses. Apparently something in the conditions of rearing is unfavourable and gradually tilts the balance in favour of the virus. Very careful attention to details of temperature, humidity, air-flow, food and other factors can, however, sometimes keep the virus inactive. There has been argument about the method of transmission: is it within the egg or on its surface? Probably both things can happen. Some congenital transmissions can be controlled if the outsides of eggs are carefully sterilized with disinfectants so that the little larva cannot contaminate itself as it breaks its way out. On the other hand some viruses seem to be carried within the egg; if larvae and other stages of an insect, it is argued, can carry latent virus within them, why not the egg? The most convincing argument is that young larvae have been observed to die of a polyhedral disease within forty-eight hours of hatching or even before they have left the egg-shell.

Infected females of a hymenopteron, the sawfly *Diprion hercyniae*, are reported to transmit virus to some, but not all of their offspring; the dose these receive determines the mortality rate and the stage of development at which they die.

Our final congenital infection is one of great practical importance. German measles or rubella is, in the ordinary way, one of the least troublesome of the infectious fevers. When, however, it affects women in the first four months of pregnancy, their babies are liable to be born with various disturbing congenital defects. A difficulty in studying rubella lies in the fact that the rash is not very distinctive: other infections, including those due to adenoviruses and echoviruses, also dengue, may cause confusion. This difficulty in diagnosis probably accounts for the belief that rubella does not lead to permanent immunity as measles does: most probably one or other of the ostensibly repeated rubella-rashes was not rubella at all. The infection may also occur in the absence of any rash. It seems to be less contagious than measles: only forty per cent of intimately exposed children between five and nine caught the infection. Only 5·5 per cent of pregnant women with no rubella-history became infected after exposure, but of course many of them may have already had the disease without knowing it.

During the years when one virus after another was being grown in tissue-cultures, rubella long remained elusive. In 1961, however, success was achieved independently by two groups of workers, Drs Parkman and Buescher [123] and their colleagues in Washington and Drs Weller and Neva in Boston [180]. A virus was grown in Boston in human amnion cells, in which destructive changes developed, though rather slowly at first. The Washington workers grew their virus in kidney-cells from African monkeys. They did not, however, observe destructive changes, but were able to detect it using the interference test, previously used in the early successful cultivation of rhinoviruses. The cultures previously infected with rubella failed to support the growth of several easily recognized viruses subsequently added. Most workers have found this indirect test for virus easier to handle than the direct method. A recent report, however, raises hopes that a line of rabbit cells may show destructive effects from rubella virus and so be the most useful yet found [102]. Anyway, there is at last available a choice of techniques for recognizing and measuring the virus and the antibodies against it.

The first suggestion of an association between rubella and congenital defects in babies came from Dr N. S. Gregg in Australia, who observed a number of cases of congenital cataract following an outbreak of rubella; heart-defects were also observed [67]. Another Australian, Dr C. Swan, soon added more information [164], and after a lot of initial scepticism the association is now recognized. The danger is only in the first four months of pregnancy when the embryo is in the early stages of development, and the earlier the pregnancy the greater the danger that serious defects will appear. One set of figures from Australia gives a rate of eighty-three per cent of defects when rubella occurred in the first month after conception, falling to sixty-one per cent for the fourth month. Some figures from elsewhere are considerably lower, but assessments have been made when the babies were of different ages. Some defects, of hearing or in the teeth, do not become evident at once. A few cases are on record of trouble when the mother's rubella had preceded conception for a short time; but rubella virus is known to be able to persist in the throat for as long as fourteen days after the first appearance of a rash, so virus may still have been around when conception occurred. Besides giving rise to babies with abnormalities, rubella may give rise to miscarriages. Australian studies yielded the following figures as to relative frequency of various defects. The commonest was deafness, with consequent inability to talk—in seventy-five per cent. Imperfect development of the brain was next commonest—seventy per cent; then eye defects, especially cataract—sixty-six per cent; malformations of the heart—sixty-three per cent; and defects in the teeth, often of major character—forty-five per cent.

The list is a formidable one, and one naturally asks: what can be done about it? Four things are possible. Rubella is such a mild disease in the absence of pregnancy that care could be taken that girls were well exposed to the disease before marriage; they should then be safe. Better still, they could be vaccinated in childhood. A rubella vaccine is not available at the moment, but research is pressing on so well that one has every hope that a vaccine will be available before long. It should be more certain and less unpleasant than exposure to infection. One may be too late, however: a woman may be exposed to rubella when already pregnant. One can then give antibody in the form of gamma-globulin from immune people in hope of preventing the disease from developing: this method is not very reliable. If rubella does develop the question arises of the justifiability of abortion. In some countries, for instance Sweden, occurrence of rubella during the dangerous first four months is considered an adequate reason for this course. Elsewhere, discussion goes on: where a young woman has every prospect of producing other children not subject to the risk of being born deaf-and-dumb, blind or enfeebled, abortion is thought reasonable. An older woman who at last conceives after a long barren period may think differently.

In the last few years rubella virus has been recovered not only from foetuses from mothers with the disease, but also from babies born with congenital defects [180]. In one instance virus was present in the throat for eight months. This might have been explained as due to immunological tolerance. Unfortunately for that theory the child's blood contained plenty of antibodies; so the finding is rather odd and unexpected. We may be seeing here another device whereby the virus survives when immediate person-to-person spread fails.

Possibly rubella is dangerous partly because it is such a mild disease: a more serious one might hit harder and end the pregnancy completely. Much study is devoted to discovering whether other virus infections can act as rubella does: there is a suggestion that abnormalities are a little higher than normal after influenza, but at present only rubella has to be taken seriously. There is, however, evidence that, in pigs, attenuated strains of swine-fever virus used for vaccination may lead to congenital abnormalities if given early in pregnancy; blue-tongue vaccines may carry a similar danger if given to pregnant ewes.

A question arises: why does rubella have this effect and during that particular period? Many viruses do more damage the younger the host they infect; but as regards rubella there is a further clue involving, once again, interferon. Dr Isaacs, working with Dr Baron [13], an American visitor to his Mill Hill laboratory, studied the effectiveness of interferon in chick embryos at various stages of development. They were aware of what rubella did to human embryos, so they predicted that interferon would work very poorly in chicks during the first third of their embryonic development; and so it turned out: the older the embryo the better the production of interferon. It is thus a reasonable hypothesis that rubella does its worst damage early in pregnancy just because the embryo's cells have not at that time acquired the ability to defend themselves by producing interferon.

Chapter 21

Viruses and Cancer: Fowl Tumours

A DISCUSSION of the relationship of viruses to cancer presents the writer with a difficult problem. So far in this book we have considered the ecology of viruses as we know about it from intelligent observation. Of viruses in relation to cancer we know almost nothing as a result of mere observation: almost everything has been learnt by means of carefully devised experiments, often under rather artificial conditions. The possibility that viruses play a part in the causation of cancer in general is, however, so important that we cannot ignore it; we must therefore consider the experimental side of the subject in more detail than is to be expected in a book on natural history.

Many viruses, those of the poxvirus family for instance, cause proliferation of cells; such proliferation is, however, limited in extent and duration and is normally followed by regression or necrosis. In cancers we see a process of cell-proliferation which is in typical cases progressive, invasive, detrimental and often fatal to the host. There are, however, gradations between the two kinds of proliferation and there are instances, perhaps not very numerous, when it is hard to say whether we are dealing with a cancer or with something else.

Much in this field is uncertain, but of a few things we can be sure. First, some true viruses are essentially involved in causing true tumours. It is possible but not necessarily true that viruses are concerned in tumours in general, including those of man. Second, no one class of viruses is alone concerned as a class of 'tumour viruses'. Those involved are some of them deoxyviruses and of more than one family; others are riboviruses. It is, however, true that some families of viruses are more likely to cause tumours than are others. Thirdly, production of tumours is an incidental, almost accidental, result of a particular kind of relation between a virus and a parasitized cell. It is of no value to the virus. Cancer commonly afflicts animals, including man, after the most active reproductive period has passed; therefore the forces of natural selection have not had as good a chance to eliminate it as if it had been a disease of early life. Its victims have already passed on their characters to the next generation before they are smitten. Fourthly, the fact that viruses can cause cancer does not mean that cancer is an infectious disease as ordinarily understood. People used to talk about 'cancer houses' in which a succession of occupants had died of the disease: this is now accepted as being nonsense; cancer is such a common condition that such things could happen as a result of mere coincidence. We have seen in this book one instance after another where a virus infection, otherwise quite latent, has been activated by some stimulus and evident disease has resulted. This is pre-eminently the case with virus-caused tumours. In almost every known instance the virus-infection is widespread but inapparent; some other factor has to come in to involve the virus in the series of events culminating in cancer.

A number of agents other than viruses are described as causes of cancer—chemicals, particularly hydrocarbons, and physical agents such as X-rays and other forms of radiation. Work has been published showing how such agents may seem to act synergistically with viruses, tumours being produced by a combination of the two, where either alone was inactive or worked much more slowly. One hypothesis is that viruses are concerned rather generally in tumour-causation, the physical and chemical agents merely serving to activate them. Another view is that viruses, chemicals and radiation are alternative 'carcinogenic' agents, operating by way of a possibly similar biochemical lesion of the cell to produce the same end-result.

Facts have lately come to light, making it extremely difficult to disentangle the relation between viruses and cancer. Viruses have been isolated from tumours, especially in mice, which have nothing whatever to do with causing those tumours; they are just there as 'passengers' because the tumours afforded a favourable place in which they could grow. On the other hand viruses may actually be concerned in causing tumours and may then disappear, at least in infectious form, while the tumour goes on growing: more about this in the next chapter.

We must now review particular examples of viruses causing cancers, and first those causing tumours and leucosis in birds. These were the first 'tumour-viruses' to be described. Ellerman and Bang in Denmark described a fowl leukaemia due to a filterable agent in 1909, and Peyton Rous in New York found a filterable fowl sarcoma soon after. This malignant connective-tissue tumour is known as the Rous sarcoma and has been studied all over the world. When it was first described, orthodox pathologists, perhaps not surprisingly, refused to take it seriously as its existence upset all their ideas about cancer-causation. Some said it was not a true virus but something else; others admitted it was a virus but said the growths were not true tumours. Both these arguments have now passed into oblivion and all agree that the Rous virus causes true sarcomas. Rous subsequently described other fowl tumours with different histological structure and many others have been reported. They appear sporadically as do cancers in other species; there is no evidence of contact-transmission.

Far commoner than fowl-tumours are cases of leucosis among fowls. Leukaemia in man and animals is a malignant disease of blood-forming tissues. Primitive cells, precursors of polymorph leucocytes or lymphocytes are liberated in excess into the blood. The cells which have become malignant are not those of fixed tissues; so we do not get a solid tumour, or not only solid tumours; the malignant process becomes diffused throughout the body. Sometimes an excess of cells is not seen in the blood and the term 'leucosis' may be preferred to describe this family of related diseases.

We recognize among fowls not only sarcomas but visceral leucosis or lymphomatosis with excess of lymphocytes and masses of tumour-tissue in liver and spleen; myelomatosis, with very many primitive polymorph-precursors in the blood; erythroblastosis, where the blood is full of primitive red cells, and finally neurolymphomatosis or fowl paralysis with nerves thickened by lymphocytic accumulations [19]. At last, more than fifty years after the early work of Ellerman and Bang and of Rous, we are just beginning to get these conditions disentangled. Fowl paralysis seems to be distinct from the rest and we will come to that later. The agents of the others are closely related to each other, both antigenically and in other ways; they make up what is called the fowl-leucosis complex.

There is a curious anomaly in that visceral leucosis, which is a widespread and economically very important disease, was for long considered as not directly transmissible, while the other leucoses and tumours have a relatively low natural incidence, yet are easily transferred by injection.

Dr Burmester of the East Lansing research laboratory, Michigan, has put forward a hypothesis which seems to cover the known facts [30]. According to this, visceral leucosis is the common, naturally occurring 'wild type' of the disease. The causative virus mutates rather easily, giving rise to the other forms of the disease, which are not contagious and are 'dead-ends' as regards natural transmission. In favour of the essential unity of the leucosis-complex is the fact that while laboratory-transmitted infections usually breed true, they do not always do so: one leucosis virus may occasionally give rise in transmission experiments to another type or to a sarcoma; and so on. The strains of myeloblastosis, erythroblastosis and sarcoma viruses most studied are in fact 'fixed viruses' whose properties have been modified and fixed by many passages in series, so that they no longer exactly resemble the original 'wild' virus.

It is now accepted that, contrary to earlier beliefs, visceral leucosis is readily transmissible: earlier failures can largely be accounted for by not having susceptible birds available for test. It can definitely be transmitted in the egg and probably also in the sperm of an infected cock; it is also quite infectious by contact. Normally, it is an inapparent infection, only a proportion of the birds dying of lymphomatosis. Enough of them, however, do die to make it probably the most serious disease for the poultry industry: an estimate indicates that a third to half of all deaths among pullets and hens can be attributed to diseases of this family. Moreover, its incidence seems to be increasing. Birds suffering from the infection, whether obviously or inapparently, shed virus in their saliva and to a lesser extent in the faeces and they continue to do this over long periods. When the infection has been acquired congenitally, the virus multiplies in the egg during incubation and chicks after hatching become a source of danger to others from a clean stock. A common drinking-fountain in a brooder unit is of particular importance in favouring cross-infection. The East Lansing research station has managed to build up a flock of birds free from the virus; few other flocks in the world are wholly free. This is of importance as a source of birds for research and ultimately, one hopes, for commercial breeding. A virus-free stock has lately become very important for another reason: some vaccines used in man are made from living attenuated viruses grown in hens' eggs. These could, and in fact often do, contain living leucosis virus. This is not known to cause any disease in man, but, for reasons which will appear before the end of this chapter, the danger cannot be ignored: accordingly every effort is now being made to see that vaccines for human use do not any longer contain this virus.

A striking difference between the malignant connective-tissue tumours, the sarcomas, of fowls and of mammals is that, in general, the former are filterable, that is can be transmitted by a cell-free filtrate, while the latter can only be passed on by injecting intact living tumour-cells. The distinction is not an absolute one, for the fowl tumours may, under several different sorts of circumstances, be non-filterable, thus behaving like mammalian tumours. Two aspects of work with fowl tumours illustrate this point. The Rous sarcoma when first discovered was only transmissible to close relations of the original Plymouth Rock hen, then to other Plymouth Rocks, then to other hens. It was found possible, many years later, to pass it on to pheasants, turkeys and ducks. The duck-adapted virus went so well in ducks that it was with difficulty readapted to fowls. Many workers had a great surprise when a particular strain of the virus was adapted to mammals, producing tumours in rats, mice, hamsters, later in ferrets and even baby monkeys [1, 2]. It would be rash to deny the possibility that it would also infect man. This is one reason why the leucosis viruses have to be kept out of human vaccines even though it is true that the mammal-adapted strain of Rous virus is a very artificial laboratory product and the risk is therefore infinitesimal. This is particularly so as the rat- or hamster-adapted tumours are found to be non-filterable, like other mammalian tumours. Suspensions of their cells can, however, be put back into fowls and will cause tumours in them; and such tumours have regained their filterability. We do not know why no free virus can be extracted from the mammalian tumours. It may prove to be explicable in the light of the next odd finding to be described, or by some of the facts recorded in the following chapter concerning 'disappearance' of viruses from tumours.

Drs Rubin [140] and Hanafusa [72] in California came across a very strange phenomenon. When the Rous virus is added to a suitable tissue-culture system, isolated foci of proliferating cells are produced. When very dilute virus is used such foci can be initiated by single virus particles. When a particular strain of Rous virus was used, some of these foci produced sarcoma cells which were fully malignant and able to induce tumours in hens but yielded no free virus. It was finally discovered that the virus of this strain was defective; it could not multiply and liberate more virus on its own but only when a helper virus was added. Normally both Rous virus and helper virus had been present, so the tumour was a filterable one. The helper virus turned out to be none other than an avirulent strain of the visceral leucosis virus which we have previously considered. It apparently acts by furnishing a coat which completes the previously defective virus, enabling it to act once more as a perfect virus, able to come into the open instead of skulking shamefully within a cell. Naturally, work is going on to see whether there are helper viruses which could turn a non-filterable mammalian tumour into a filterable one. Instances of completion of defective viruses by a helper virus are also on record among plant-viruses and bacteriophages.

It is worth noting that the enormous mass of remarkable knowledge which has been acquired during fifty years of study in this field has all come from work on the tumour-like conditions in a single species of bird, the domestic fowl. Though the viruses have, as we have seen, been transmitted to other bird species, naturally occurring filterable tumours have not turned up spontaneously in any of these. It is true that fowls have been far more studied than other species, but it is also true that fowls have, over hundreds of years, been modified by selective breeding so that they represent rather an unnatural artificial kind of creature.

Finally, a word about neurolymphomatosis, otherwise known as fowl paralysis or Marek's disease. This has been extremely prevalent and second in its importance to the poultry industry only to visceral leucosis. It has been repeatedly stated that it is not transmissible by injection, but such transmission has now been achieved by Biggs and Payne working at a research station in Huntingdonshire [20]. The matter was rendered obscure by two facts: transmission was never successful to more than a certain proportion of birds, and then no flock has been actively free from spontaneous Marek's disease, so that some cases were always turning up in control, uninoculated birds. One had always to depend, therefore, on a percentage difference between incidence in inoculated and control fowls: not a very satisfactory matter. Though no completely Marek-free flocks are known, that used by Biggs and Payne had such an extremely low incidence that experiments could be interpreted with safety. The agent recovered has been passed through filters, and it is antigenically unrelated to those causing the other kinds of leucosis. In fact, no antibody-production against the Marek agent has been detected at all. It has not been cultivated, is very difficult to study and we do not yet know whether it is a very distant relation of the other leucosis viruses or belongs to quite a different family. It appears, like the others, to be transmitted through the egg. In the common form of the disease paralysis of limbs follows great thickening of the nerves as a result of infiltration by lymphocytes; another form is known as 'grey eye' from changes occurring in the iris. It should not be regarded as a tumour-like condition in the usual sense but is best considered along with the other leucoses. While visceral leucosis has for some years been on the increase, fowl paralysis has fortunately been on the decline. Unfortunately a very acute form of what seems to be this disease has been turning up recently; in this the lymphocytic accumulations are in the internal organs rather than the nerves—another cause of confusion with other types of leucosis. It will by now be crystal clear that this whole subject is a very obscure and difficult one.

Chapter 22

Virus-tumours in Mammals – and Frogs

IN THE LARGE MAJORITY of cancers in mammals, including man, no virus can be demonstrated. Is this because viruses are not concerned, or because they are occult or elusive? Much current work on cancer is directed to finding an answer to this question, and naturally particular attention has been paid to the exceptional tumours which do seem to have a virus as an important link in the chain of causation. Note this phraseology: 'an important link' not 'the cause'. Findings with these tumours suggest possible solutions to puzzles concerning mammalian cancers in general.

The papillomas or warts, many of which are caused by viruses, stand apart from the other tumours to be considered. They are benign tumours, overgrowths of cells of a particular kind, often persisting for a long time, but growing in an orderly way and usually regressing in the end. A number of wart-viruses are known, infecting man, cattle, horses, goats, dogs, rabbits and other species. One, which affects cottontail rabbits in the USA (plate 39), was discovered by R. E. Shope and has been studied in detail by him and by the pioneer of work on virus-tumours, Peyton Rous. Like other wart-viruses it is probably transmitted by direct contact. Bits of wart can easily be rubbed off and the virus is a very stable one. Biting-insects can transmit infection experimentally, but their role in nature is very doubtful. The infection can be passed from the cottontails (*Sylvilagus*) to domestic rabbits (*Oryctolagus*) by rubbing warty material into lightly scarified skin: and here the matter begins to get more interesting. For, though the warts take readily in tame rabbits, it is very difficult, often impossible, to recover infectious virus from them. All the same, something of a viral nature must be present in these warts, for extracts of them will immunize rabbits against infection by active cottontail wart material. There has been dispute as to how these non-infectious domestic rabbit warts differ from the infectious ones from cottontails: is the difference qualitative or merely one of quantity? Cotton-tail warts have been transmitted by means of DNA extracted from them: Shope has suggested that in the *Oryctolagus* warts, the DNA is present, while little is produced of the protein which coats the DNA. In fact papillomas have been produced with DNA extracted from tame rabbit warts.

Another matter of interest now crops up. Though the tame rabbit warts usually regress, they not uncommonly show instead a malignant change and at one or more points on them an invasive cancer develops, capable of ultimately killing the rabbit (plate 40). This sometimes happens with the warts of cottontails, but much more rarely. As a rule, no papilloma virus can be recovered from the cancers, yet, as with the papillomas from which they arose, virus-material can be detected indirectly through the ability of extracts to induce immunity. In one line of derived cancers, however, even this property has been lost and the growths behave just like any non-filterable mammalian cancer. Have the virus and its products finally disappeared, or has something from the virus become so closely integrated with the nucleus of the cell that no immunizing material is liberated any more?

The rabbit papilloma has proved to be of interest in yet another way. Rous and his collaborators have shown that the virus may act synergically with chemical carcinogens in the production of cancer. When the virus was given intravenously into rabbits whose skin had been previously treated with tar, it localized in the treated areas and moreover gave rise, not to the usual uniform, orderly growths, but from the very beginning to bizarre growths of various types, many of them malignant from an early stage [135]. In another type of experiment, areas of skin inoculated with the virus were covered with a dressing containing a carcinogenic hydrocarbon [94]. The papillomas which arose were rather poor ones, yet many of them gave rise to cancers much more quickly than happens after administering either virus or chemical by itself.

It was mentioned above that papilloma-infection was probably contracted by rubbing off bits of infected wart tissue; so the wart itself helped in the propagation of the virus. The virus-tumours next to be considered seem to be by-products of a virus's activity, in no way of value for its survival. When controversy was still active as to whether the Rous sarcoma was a true tumour caused by a true virus, pathologists were startled by reports from Dr John Bittner that, in mice, cancer of the breast, that most classical of all tumours, had yielded a causative agent [21]. This was first described as an extra-chromosomal factor, that is something handed down from one generation to the next otherwise than in the genetic apparatus of the chromosomes. It turned out to be passed from mother mouse to her offspring in the act of suckling. Inbred strains of mice were in existence, some of them showing a natural incidence of breast cancer of ninety per cent, others with virtually none. If newborn mice of the latter kind of strain were removed from their mothers at birth and suckled on mothers of a high-cancer strain, many of them developed cancer six months to two years later. Conversely if baby mice of a high-cancer strain were fostered by low-cancer mothers, few of them developed the disease. The operative agent in the milk has been referred to as a 'milk factor', an 'influence' and so on; but all pretence that it is not a virus has now been abandoned. It is probably an RNA virus, rather like a myxovirus to look at, though we do not know whether the structure of its nucleocapsid is helical, as in myxoviruses.

The role of the virus in causing cancer has, however, been complex and hard to sort out. A number of factors operate in deciding whether a mouse develops breast-cancer or not: it is favoured by presence of the virus, by a genetic background of susceptibility and by hormonal factors. In female mice, a rapid succession of pregnancies favours; in males, the injection of certain hormones. Other significant factors are temperature, diet and over-crowding. Where there are multiple factors such as these in causation, an excessive quantity of one of them may make it appear that this is *the* cause. On the other hand, one may doubt their causative role when tumour-viruses given to an animal in large doses may seem merely to accelerate the appearance of a tumour which would ultimately have turned up anyway.

The Bittner virus is most effective if given to very young mice; but very big doses will overcome the resistance of older ones. It is present in the blood of infected mice and especially in apparently normal mammary glands, even more than when these have developed tumours. Mice which are infected may, if they are lucky, never develop tumours; nevertheless they go on carrying the virus and pass it on in their milk to their offspring. If, however, the mice are of a relatively resistant strain, accepting the virus with reluctance, then it may gradually die out, being extinguished in the course of two or three generations. The virus may be present in wild mice as well as in tame ones. The condition has been difficult to study as different workers have obtained very conflicting results. One difficulty is that particles just like those of the Bittner virus have been found in the breasts of normal mice, supposedly free from the agent. Another is that serially transplanted tumours originally induced by the virus, may in the course of time cease to yield any: a result recalling that noted with the rabbit warts and cancers derived from them.

Another tumour occurring spontaneously in some strains of inbred mice is lymphatic leukaemia; this may present itself as a typical leukaemia with many abnormal cells in the blood, or as a lymphosarcoma, a malignant growth in one or more organs, especially the thymus gland. Dr Ludwik Gross in New York studied a strain of mice with a high incidence of leukaemia and found that he could transmit a filterable agent from their leukaemic tissues to newborn mice of a strain in which leukaemia rarely occurred naturally. His results were received with much scepticism, but they have now been amply confirmed. The leukaemia virus is transmitted from mother to offspring—and for this Gross coined the term 'vertical transmission'; this may be through the milk, as with Bittner's virus; but in general the embryos of 'carrier' mothers are already infected when born [68, 69]. The virus contains RNA and is very similar, by electron-microscopy, to the Bittner virus. It may be hard to find virus in naturally infected mice, and leukaemia in other mice injected with such 'wild' virus turns up irregularly and only after a long incubation period. However, serial passage through baby mice has led to the production of a 'hotted-up' virus which is easier to study. It can give rise to leukaemia only two to three months after injection, it can infect rather older mice, not only newborns, and also baby rats: it usually produces lymphosarcomas and lymphatic leukaemias but may also give rise at times to other tumours or leukaemias involving other cell-types. In this respect it resembles the fowl leucosis virus. It seems, by the way, merely a matter of convention to talk about the fowl 'leucosis' complex and yet to call a similar complex of diseases among mice 'leukaemia'.

So far as is known, natural transmission of the infection is always vertical, never by contact. As in the case of the Bittner virus, infection is not necessarily followed by development of the disease: hormonal and other external factors may help to decide the outcome. There is, once more, a relation to other non-viral carcinogenic agents. Several groups of workers have produced leukaemia by irradiation in mice not known to be infected with the virus and yet have recovered the agent from the leukaemias which arose. It seems that the irradiation must have activated a latent virus. Among the tumours produced by Gross with leukaemia material were cancers of the parotid salivary glands. Their occurrence confused the work prior to 1958, when it was discovered that quite a different agent, the polyoma virus, which we shall be dealing with shortly, was present as a contaminant in leukaemia material and was producing the parotid growths.

In considering the fourteen or more other leukaemia viruses which have been discovered in mice we are indeed getting very far from 'natural' history, for these viruses have not come from spontaneously occurring leukaemia in mice, but mostly from other kinds of mouse tumours with the causation of which they had almost certainly nothing whatever to do. Leukaemia has appeared during serial propagation of these other tumours, and an agent has been recovered which thereafter bred true as a leukaemia virus. It seems fairly certain that these viruses are latent in certain stocks of mice, producing in them no evident disease at all. They apparently find in the laboratory workers' transplanted tumours a very favourable habitat; in this they multiply to an abnormal extent but are mere passengers, yet are ready to show pathogenic powers when inoculated into another animal. They differ from each other in a number of biological characters and it is hard to say how many really different viruses are present in the collection. They are hard to sort out by antigenic tests; certainly they are not all distinct in this respect. They seem to be RNA viruses and very similar to both the Bittner and Gross viruses (plate 41). Electron-microscopy sometimes shows that they have tails and look like tadpoles, but most workers think the tails are artefacts produced in the course of making preparations for examination [45]. It seems likely that the mouse-leukaemia viruses are members of a family of related viruses, of which all have the potentiality of producing leukaemia, and that one member, Gross's virus, has the power of doing so naturally in a particular strain of inbred mice.

There is evidence that a leukaemia in cats is caused by a virus: there is a possibility that leukaemia in cattle may be similarly caused; and a great deal of work is going on by people trying to discover if the same is true of leukaemia in man. Several claims to have isolated a human leukaemia virus have been made but none has been substantiated: some of the 'viruses' have turned out to be very tiny organisms of the genus *Mycoplasma*, only cultivable in very special media.

The 'tumour-viruses' of mice so far discussed have all been riboviruses; the other viruses to be considered in this chapter are all deoxyviruses, and we come first to the polyoma virus already mentioned [52]. It is present as an inapparent infection in many stocks of mice and produces tumours on inoculation in large doses into newborn mice or hamsters: rats, guinea pigs and ferrets are also susceptible. In small doses or when given to older mice it causes an inapparent infection. The suffix '-oma' is used by pathologists to describe tumours of various kinds of tissues—carcinoma, sarcoma and so on. The tumours produced by this virus are of such diverse nature that the term 'polyoma' was coined, implying that 'many' kinds of growth could be caused. The virus belongs to the same family of viruses as those causing papillomas. It may produce destructive changes in some kinds of tissue-culture but in others it causes heaping-up of cells and cellular changes, so-called transformations, which represent a partial or complete conversion into cancer cells. The mechanism of the change has been the subject of much study. The tumours produced may be transplanted in series by means of cell-grafts and when this is done the virus may seem to disappear from the scene. It was discovered by Dr K. Habel [70] that when mice were injected with polyoma virus under conditions where tumours were not produced, the mice became immune to grafts of the transplanted polyoma tumours. This was very surprising, because immunity to a virus and to grafted cells are normally quite separate phenomena. Dr Habel then showed that the polyoma tumours had developed a new antigen, distinct from the ordinary polyoma antigens and that this was responsible for the transplantation-immunity: growth of the virus given as an immunizing dose can be supposed to generate some of the new antigen, even though tumours do not develop; it then immunizes the animal and renders it able to reject the tumour-grafts. Here is more evidence that a virus can play a specific role in causing cancers and yet fail to be recovered, at least in a complete form, from the growths it has caused.

We do know something of the natural history of polyoma virus in mice. It is present in wild mice both in urban and rural communities [84]. It is not known ever to cause cancers in such animals; perhaps they do not receive big enough doses or perhaps they do not live long enough. It is excreted in large amounts in the urine and also in the faeces; their bedding may be freely contaminated. It is infectious when taken in by mouth. It is likely that grain used for human consumption may often contain the virus, but there is no evidence that it can infect man. This is a very stable agent and may exist in the environment of a laboratory, getting into places where it is not wanted and leading workers to all sorts of erroneous conclusions.

Another virus of the same family as polyoma is the vacuolating virus of monkeys, so-called because of the vacuoles it produces in cultivated cells. It is commonly known as simian virus 40 (SV 40) [165]. It has been found to be present in batches of polio vaccine, but was long undetected because it does not readily produce any changes in rhesus-monkey cultures used for making those vaccines. It does, however, destroy cells in cultures of kidneys of African monkeys and use of such cultures permits its detection. It will multiply in man, but has not been discovered to produce any harmful effects. When, however, cultures are injected into baby hamsters, tumours may be produced. Just as in the case of polyoma virus, the SV 40 usually disappears from the cultures and a new antigen, particularly associated with these tumours, turns up. In tissue cultures of several species, including man, this virus produces cell transformation such as was alluded to in discussing polyoma; but in cultures of tissues of monkeys, the natural host, this does not happen. People may well ask, 'Isn't it a terrible thing to inject people inadvertently with leucosis virus and SV 40 which can produce cancer in animals?' The answer is that it is most undesirable and now that the danger is recognized it should not happen any more. On the other hand the yellow-fever and polio vaccines in which these agents have turned up are extremely effective, have saved very many people from death or crippling and the danger from the 'accidental' virus is purely hypothetical.

We get a little nearer to possible human cancer viruses in considering the next group, the adenoviruses. Many of these infect man, causing, as mentioned in chapter 5, sore throats and other symptoms. At least four of the serological types will produce tumours in that very susceptible creature, the newborn hamster; and tumours have also been caused in rats and mice. Once more, we find that the virus 'disappears' and that new antigens are to be recognized. This time the story is a little different in that, at least in the case of type 12 adenovirus, the tumours which contain no infectious virus nevertheless contain a specific antigen reacting in the complement-fixation test with antibodies against the virus itself. Also 'disorganized' virus particles have been recognized in such tumours by the aid of electron-microscopy. It seems that an incomplete virus, or part of a virus, can be carried on in multiplying tumour cells, unable to get out and infect new cells and possibly, though we do not know this for certain, actually cause the malignant behaviour of the cells. In some parts of the world, particularly around the Indian ocean, cancers are common in the pharynx, a region favoured by adenovirus infections; work is going on to see whether or not adenoviruses may not be concerned in the causation of these cancers.

Two other families of viruses have to be considered in relation to tumours. Many poxviruses cause proliferative lesions which just fall short of being considered as true tumours. We have already considered the fibromas of rabbits; and there is the Yaba virus causing tumour-like growths in Asiatic monkeys housed in the open in West Africa [120]. This may well have been carried by insects from a naturally infected African primate or other animal.

Then there is a virus of the herpesvirus family found in kidney-cancers in leopard-frogs (*Rana pipiens*) [101, 131]. These growths occur in about two per cent of frogs of this common North American species, especially in Vermont and adjacent parts of Canada. The tumours contain large intranuclear inclusion bodies of a type characteristic of the herpesvirus family. A virus, looking like that of herpes simplex, can be isolated; by whatever route it is inoculated into a frog it is in the kidneys that the growths appear after some months; the virus seems to localize only there. It is suggested that injection of the virus only accelerates the appearance of growths which would have turned up anyway; but we have already seen that that seems to be happening with other tumour viruses. The virus has produced bony growths when injected into young salamanders.

There is one cancer of man for which a virus origin has been very strongly suspected; and search for that hypothetical virus has been very intense. A very malignant lymphomatous tumour, called the Burkitt tumour after its

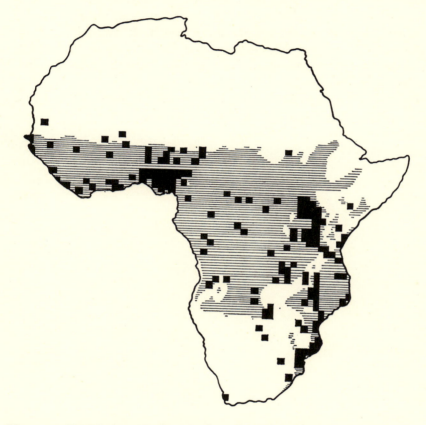

Figure 22. The black squares show where Burkitt tumour has been found: the thatched areas show regions of higher temperature and rainfall, where mosquitoes are especially prevalent.

original describer, occurs in parts of Africa and affects particularly the jaws of young children [29]. Its incidence is related to physical conditions in the environment, particularly height above sea level, in a manner which is correlated with the activity of mosquitoes; so a mosquito vector is strongly suspected. Several claims to have isolated a virus have been made, and indeed viruses have been recovered, particularly herpes and reoviruses, but these are probably 'passenger' viruses such as have repeatedly turned up in work with mice. A convincing viral cause has still to be produced. It may be noted that some objects looking very like herpesviruses have been seen in material from which no active virus can be extracted [56].

The tumours described in this chapter are perhaps exceptional cases among the vast numbers of growths known to occur in all sorts of species. There is no certainty that viruses are concerned in most tumours. On the other hand, we have seen that viruses can be very much involved in causing tumours, yet may seem to disappear as such from the growths which result; their involvement may still be recognizable by indirect tests such as serology. Also, tumours apparently caused by chemical or physical agents may nevertheless have viruses present as links in the chain of causation. We have therefore to keep a very open mind on the whole matter.

Chapter 23

Not Really Viruses

THIS CHAPTER will consider organisms related to psittacosis, agents which were for a long time considered as viruses but are no longer included in the category. Rather they are to be classified with the rickettsiae, small bacteria which are intracellular parasites; these, too, have often been dealt with alongside viruses in textbooks because of certain similarities in behaviour. These organisms deserve our attention because their natural history is often so similar to that of viruses. It can thus be made clear that most of the general principles discussed in this book apply not merely to viruses but to parasites in general.

The agent which causes psittacosis or parrot-fever is a member of a rather tidy compact family of organisms, all of them having an antigen in common, though of course they are not antigenically identical. Other members of the family cause lymphogranuloma venereum (LGV), a venereal disease of man, pneumonia in cats, mice, sheep and goats, abortion in sheep, encephalitis in cattle and human eye infections—trachoma and inclusion conjunctivitis. We shall have more to say about some of these others when we have dealt with psittacosis. There is unfortunately no convenient name for the family: they have been called the P-LV group and other names based on the initials of prominent members. Being bacteria, they come under the *Bacteriological Code of Nomenclature* and belong to the Chlamydozoaceae, most of them falling into the genus *Miyagawanella* and those infecting eyes into *Chlamydozoon*. These are clumsy names and some prefer *Bedsonia* for the psittacosis-like members, after S.P. Bedson who contributed greatly to early knowledge about the family. This is a euphonious and in many ways reasonable name: unfortunately the clumsier names have technical priority and have to stand unless set aside by official action.

The reasons for ejecting these organisms from among the viruses are as follows. Unlike viruses, they contain both RNA and DNA; they have cell-walls containing muramic acid, a substance generally present in the walls of bacteria; there is evidence that at one stage in development increase occurs by binary fission, splitting into two. Then their metabolism must resemble that of bacteria, for they are susceptible to the action of certain antibiotics, particularly tetracyclines, which are active against many bacteria. On the whole they are rather larger than large viruses, at least at some stage in development.

Some very important members of the family are parasites of birds. Those infecting members of the parrot family, the psittacines, have attracted most attention because they have been particularly apt to spread from their natural hosts and cause disease in man. The disease so caused is often a pneumonia and this is apt to be serious in older people: it is known as psittacosis. It is illogical but convenient to lump together all the infections contracted from other bird species as 'ornithosis'.

Psittacosis affords an excellent example of an infection which is normally latent but may be activated under conditions of stress; such conditions have been well studied only on parrots in captivity, particularly in breeding of colonies of *Melopsittacus*; these are often known in America as parakeets and elsewhere as budgerigars. Psittacosis is endemic among numerous Australian and South American psittacines in the wild; it rarely causes recognized disease, though if it does so among nestlings, this would probably not be known. The agent was recovered from 27 out of 206 wild birds of the parrot family caught in Australia in 1935; the positives were in five different ones of the thirteen species examined. Exceptionally the agent may cause epizootics in south-eastern Australia and Tasmania. In an outbreak in 1938–9 it was reported that a 'large number of wild parrots . . . have been dying during the past month (October). Literally hundreds have dropped and are lying about in paddocks, found caught up in trees etc. Other birds are not affected.'

This, however, is not the usual story: psittacosis and related agents commonly operate most stealthily. Karl Meyer studied in California what happens among breeding budgerigars [115]. Young birds in their first season often carry virus but remain healthy and do not excrete it. Under the stress of nesting, the host-parasite balance was upset, virus was again shed and infected the young birds in the nest. A few – up to ten per cent – died, but the rest appeared healthy, though many continued to carry virus until in their turn some factor upset the balance. Some birds may continue to carry virus for life but many become immune and the infection ends. Miles, working in Australia, suggests for this reason that silent carrying of virus cannot perpetuate infection among wild parrots, and that under natural conditions an occasional epizootic among fledged birds is necessary if the virus is to survive indefinitely.

Under the conditions of captivity all too many other stresses affect captive budgerigars and other species. Chief among these is crowding together during shipment abroad of wild-caught birds, or attempted rearing under unhygienic conditions. When birds are reared under wholly satisfactory conditions there may be no evidence of disease at all, yet latent infection may still be present. When birds from such a source are submitted to the stress of shipping or are transferred to someone who takes less proper care of his pets, the agent may be excreted by the carriers and infect other birds. Meyer found latent infection to be present in a quarter of the flocks he examined in California. It is no easy matter to eliminate the infection from flocks and so to make them safe as pets. Fortunately it can be done by adequate, thoroughly supervised, administration of tetracycline drugs, particularly aureomycin, in the birds' food.

With infection often very prevalent among the birds, it is fortunate that human infections are relatively infrequent. At least we assume that this is so, though many people have antibodies in their sera which cannot readily be accounted for. Infection of man is probably through inhalation of dust particles contaminated by the birds' droppings. There have been occasional outbreaks of human disease. In one recorded instance twenty-six people who entered a room containing apparently healthy parrots contracted the disease and five of them died. Presumably an unusually virulent strain was present, or else there were birds shedding virus into their environment with unusual generosity. Human infections rarely occur, however, from contact with birds showing no signs of illness.

Turning to ornithosis, the infection of birds other than psittacines, we find that the turkey is of most importance, particularly in North America. The infection in birds is usually chronic, leading to droopiness, diarrhoea and other symptoms. Most human cases have occurred in people handling killed turkeys, plucking and dressing them, before despatching them to the market. Infection may also be contracted from infected chicken flocks or, in Czechoslovakia, from ducks, but spread of agents from these birds is relatively uncommon. Pigeons, on the other hand, are widely infected though, fortunately, with less virulent strains which are not very apt to infect man. The ornithosis organism has been recovered from pigeons in Trafalgar Square. Outbreaks have been reported among people plucking and cleaning Fulmar petrels which are used for food in the Faeroe islands. The agent occurring in egrets is said to be of unusually high virulence.

It is unusual for cases of psittacosis or ornithosis in man to give rise to further human cases; however short chains of man-to-man infection have been recorded. There are also records of transmission in series of infections with organisms of the family slightly different antigenically from avian strains and arising in people with no known contact with birds. It could be that the avian origin has been lost, the apparently original case having picked up infection from an unidentified person who *did* have contact with infected birds. It has also been suggested that these are infections with a member of the group essentially adapted to be a human parasite. Against this is the fact that the various instances have been rare, widely scattered, involving only short infection chains and with organisms somewhat differing from each other antigenically.

A truly human member of the family is the agent of lymphogranuloma venereum (LGV), otherwise known as tropical or climatic bubo. It is usually transmitted venereally, and is commoner where social conditions are poor, in the tropics and in larger seaports. The name tropical 'bubo' refers to the enlarged lymph-glands which are often a feature of the disease and occur particularly in the inguinal regions. We may note here that transmission of infection by the venereal route has not been mentioned in earlier chapters dealing with viruses: it is, however, of importance in some virus diseases. For instance, a member of the herpesvirus family, infectious bovine rhinotracheitis, is recognized as a respiratory infection of cattle in Britain, Germany and the western USA. There was what seemed to be quite a different disease in the eastern USA and some other countries, transmitted venereally and known as infectious vulvo-vaginitis or coital exanthema. It is now established that the same virus is responsible for all these infections, a different picture occurring according to the prevalent means of transmission.

The other agents related to psittacosis which need mention are those causing trachoma and inclusion-conjunctivitis [43]. Though still related to psittacosis antigenically, they are not so close to it as are others of the group and they have been placed in a different genus, *Chlamydozoon*. Trachoma is a chronic disease of the eyes, known since classical times and a major cause of blindness. Infection commonly begins in childhood, being often contracted from a mother; the chronic inflammation leads on to opacity of the cornea, scarring and eye deformities; secondary bacterial infections increase the damage. Its interest from the point of view of ecology lies in its distribution. Though formerly widespread, it has now virtually disappeared from European countries with a high standard of hygiene. It is very prevalent in the Middle East, particularly Egypt; also elsewhere in Asia and Africa. Eye-to-eye transmission is probably the rule but indirect transmission may occur and flies are thought to play a part. In North America the disease still persists, particularly among Indians on reservations where hygiene is poor. Trachoma is a disease associated with dirt.

Differences from the disease inclusion-conjunctivitis are of much interest.

The name derives from the presence of inclusion-bodies, which are colonies of the agent in infected cells and are like those occurring in trachoma. By most laboratory tests the agent cannot be distinguished from that of trachoma. The natural history is, however, very different. The agent normally inhabits the genital tract and babies get infected from their mothers at birth. The same may be true of a bacterial infection, gonorrhoea. Conjunctivitis caused by the inclusion-body agent is much less serious than is trachoma. It is self-limited, persisting as a rule only for a matter of months and it does not cause scarring and other associated troubles. Eye-to-eye transmission is probably uncommon. The disease when it occurs in adults may be more chronic than in babies and the inflammation is less acute. There have been reports of outbreaks of infection contracted in swimming-baths, but 'swimming-bath conjunctivitis' may also be due to adenoviruses. Inclusion-conjunctivitis can occur anywhere in the world and, in contrast to trachoma, is not particularly associated with poor hygiene.

We should be straying too far from the subject of this book to consider at any length the ecology of the small parasitic bacteria belonging to the rickettsiae. It must suffice to make brief mention of some high-spots of interest, particularly those which recall similar findings in the virus field. Mention has been made of a very speculative hypothesis concerning a possible origin for viruses in the symbiotic bacteria found in cells of insects. This may be a wild idea as regards viruses, but it is quite a reasonable one as far as rickettsiae are concerned. These are very similar in appearance and in many properties to some of the insect symbionts; one such organism, found in the sheep-ked, a parasitic fly (*Melophagus ovis*), is commonly classed as a rickettsia, differing from other rickettsiae in that it can be grown on artificial media. Where such micro-organisms, living in insects, happen to occur in blood-sucking ones, it would seem a very natural thing that in the course of evolution they should be modified in a way allowing them to use the attacked vertebrate as a means of getting from one arthropod host to another. A secondary development would be acquisition of the power to multiply in the vertebrate as well as in the arthropod.

Rickettsiae were probably all arthropod-transmitted to begin with, and this is still the standard mode of transfer today. Classical typhus is transmitted by body-lice, endemic or murine typhus by fleas, other kinds by ticks or mites. The late Hans Zinsser in a fascinating book *Rats, lice and history* [186] has discussed how epidemics of louse-borne typhus have again and again changed the course of history. He has also marshalled the evidence for the view that louse-borne typhus has been derived from murine typhus. The latter is naturally a disease of rats and mice and is transmitted by the rat-flea, *Xenopsylla cheopis*. The rickettsiae are present in the fleas' faeces and man becomes infected by rubbing in such infected material when he scratches himself.

The parasitism is doubtless an ancient affair for it normally hurts neither rat nor flea. Louse-borne typhus, on the other hand, not only causes serious illness or death to its normal host, man, but is also fatal to the lice – good evidence that the parasitism is not of remote origin. Again, man is infected by scratching in infected louse-faeces and also by inhaling dried rickettsia-containing dust. We think of the danger of contracting typhus from lice where hygiene is poor. The lice, too, could well be alive to the danger of catching the disease by biting unsatisfactory, unhealthy subjects!

There are a number of rickettsiae carried by ticks in various parts of the world. That most studied has been the Rocky Mountain fever of North America. This rickettsia can be transmitted transovarially to the next generation, and its ecology presents interesting similarities and contrasts with that of some arboviruses, especially Colorado tick fever. The two diseases may occur in the same areas. Mite-borne or scrub-typhus of the Far East was a source of much danger to our troops fighting in the jungle during the last war. The extent of the risk depended on the abundance of the rodents which are the natural hosts of the rickettsiae, and so upon the measures which favoured or discouraged the rodents. The use of anti-mite repellents and of chemotherapeutic agents finally reduced the menace to something not too alarming.

Perhaps most fascinating among the rickettsiae is the agent of Q- or Query-fever, also deserving a Q because it first turned up in Queensland [46]. Here it seems to be an infection of a marsupial, the bandicoot, and to be carried by a tick, *Haemaphysalis humerosa*. As a human infection, unpleasant but rarely fatal, it appeared among cattlemen: what apparently happens is that certain other ticks bite cattle as well as bandicoots and pick up the virus; their droppings dry up, as mentioned for other rickettsiae, and are inhaled by men handling cattle-hides. The infection was next recognized among troops in Italy, where it was apparently caught by inhaling dust in hay-lofts and such places. It has since appeared in different parts of the world, different ticks and different reservoir hosts being concerned. In places ticks may be responsible for maintaining an infection in some reservoir, but elsewhere not even this is certain. In the USA and parts of Europe, including Britain, the infection is endemic among the sheep and cattle of some districts [83]. The rickettsiae may be excreted in milk and cause infection in people who drink unpasteurized milk. It can also occur, once again, by inhalation of infected dust. Placentae and birth fluids of infected sheep and cattle are especially rich in virus and are a particular source of danger to those in contact with these animals at the time of parturition. Q-fever is known to be a dangerous organism for laboratory workers to handle, for many have become infected. More than fifty caught the disease in one institute, including people in parts of the building remote from where work on it was in progress [82]. We saw earlier how some viruses, originally arthropod-borne, had now become transmissible by other routes. This may be an important principle in understanding the evolution of viruses attacking vertebrates. The *Rickettsia burnetii* which causes Q-fever affords an admirable illustration of a versatile and adaptable parasite capable of surviving in different hosts, in different parts of the world and making use of different methods of transmission.

Chapter 24

Man and Virus Ecology

PRECEDING CHAPTERS will have made it abundantly clear that virus infections are, like other examples of parasitism, a matter of host-parasite equilibrium. With some exceptions, this 'normal' state of affairs does not lead to any manifest disease. Only when something upsets the equilibrium does disease appear. Too often man and his activities furnish the necessary unbalancing factor.

We have so far mainly considered virus ecology as a subject worthy of study for its own intrinsic interest. Viruses, however, concern man's prosperity and happiness in many ways, and this last chapter will therefore bring together the many instances, many of them alluded to in previous chapters, in which man has either favoured the viruses and the development of virus disease or has found ways of frustrating their activities for his own benefit.

Man favours viruses either deliberately or, much more often, inadvertently. He has introduced myxomatosis to control plagues of rabbits, with varying success (see p. 149). He has used insect-pathogenic viruses to combat sawfly and other caterpillars ravaging his forests or his crops (see p. 124). On the other hand he has brought his horses, his cattle, pigs or sheep into areas where there exist endemic viruses harmless to native species but dangerous, often fatal, to his herds. Examples are equine encephalomyelitis (see p. 79), malignant catarrh (see p. 147), African swine fever (see p. 147) and probably African horse-sickness (see p. 116) and blue-tongue (see p. 115). At other times he has unwittingly introduced viruses, harmless to animals in their natural home, but far otherwise in the countries to which he has brought them. Some diseases of sheep in Iceland (see p. 170) illustrate this point. Again by establishing 'pure cultures' whether in crowds of people or in herds and crops, he favours endemic viruses which might otherwise be of little importance. Alteration of the environment by clearing jungle, draining or irrigating has favoured certain viruses, particularly by encouraging reservoir hosts or vectors. Finally, we have an instance in which the very hygienic measures which have helped to control typhoid and other bacterial infections have increased the danger from paralytic polio (see p. 70).

Our main need, however, in this chapter is to see how we can learn from past mistakes to avoid the consequences of inadvertent and misguided favouritism towards viruses, to discover, if possible, how to control those which are harmful. In protecting himself, his crops and herds, man will only infrequently seriously interfere with the natural ecology of a virus or succeed in exterminating it. For, as we have seen, very many viruses exist in reservoirs in wild animals and these will remain untouched unless the wild species themselves perish altogether. The total extermination of small-pox is an avowed aim of the World Health Organization and is not beyond the realms of possibility in the foreseeable future. Vaccination with attenu-ated strains of poliomyelitis could lead to the complete replacement of viru-lent strains by harmless ones. In these two instances we have viruses which are, so far as we know, purely human parasites. Extermination of yellow fever in South America was thought to have been accomplished before the discovery that jungle fever existed in the monkeys there.

Extermination of a virus from domestic animals is a practical possibility in those countries where there is no reservoir. Thus, Britain keeps free from rabies and also, when no fresh virus is introduced from abroad, from foot-and-mouth disease. Quarantine and a slaughter policy make such things possible. Australia has thus kept a number of virus-diseases away from her shores.

On the whole, however, we have to admit that most viruses must be allowed to continue undisturbed in their natural habitat: all we can do is to see that they stay there and do not stray out and do us harm. We are then failing to affect a virus's ecology at all – merely keeping it at arm's length. From the virus's point of view we are often merely closing the entrance to a blind alley which would not have benefited the virus anyway.

Man's own point of view must not be neglected, and we will conclude by a brief review of those ways in which we can avoid harmful virus disease. First of all we can try to ensure that no contact takes place between a virus and ourselves or our susceptible flocks and herds. Anyone in his senses will do his best to keep out of the way of a mad dog; but it is not always as simple as that. It should be possible to avoid going into forests where there is danger of picking up ticks infected with, let us say, Kyasanur forest virus (see p. 110); or else to wear protective clothing or to use deterrent chemicals. Mosquitoes can be thwarted by netting, by avoiding dangerous places at dangerous times or, again, by using anti-mosquito deterrents. Those who have to handle monkeys can be trained to use methods which do not expose them to the risk of being bitten and infected with B virus. Hygienic measures may also achieve the desired ends. Stopping the spread of poliomyelitis and other enteroviruses by the faecal-oral route should not be difficult since water supplies can be rendered safe, but this will only be wholly successful if there are no alternative means of transmission available. Attempts to achieve an air-hygiene as successful as water hygiene have so far proved disappointing: often the contact between man and man is too close to give air-sterilizing measures a fair chance. Proper sterilization of syringes, however, should be eminently successful in cutting down the incidence of hepatitis; better knowledge should also reduce the danger of transmitting the same infection in the course of transfusions (see p. 169). Public health measures may be of value: regulations may be made to prevent unnecessary gathering of crowds in the face of an epidemic; tracing and isolation of contacts can prevent smallpox from getting out of hand. Similar measures, combined with slaughter, are used to control foot-and-mouth disease and other infections of livestock. We learnt in chapter 15 how differing methods of husbandry, different treatment of weeds, different timing could favour or discourage the attack of viruses upon crops. Use of seed-potatoes from the relatively virus-free areas of the north is a time-honoured method of avoiding trouble from potato-pathogens.

There remain, however, instances when virus infection cannot be dealt with merely by such avoiding actions as these. Other methods must be brought into play. Most important of these is protection of man or his domestic animals by vaccination. To prevent an infectious disease from causing serious trouble it is fortunately not necessary to immunize one hundred percent of the population at risk. The likelihood that a virus will cause a spreading epidemic depends upon the chances that an infected host will spread the disease to another host. If it spreads to more than one, the outbreak will be a developing one, if to less than one it will be on the wane. It wanes anyway when a sufficient number of persons or animals have been rendered immune. If vaccination has immunized a large proportion it will wane all the sooner or the outbreak will not develop at all. Living attenuated virus-vaccines commonly give much more lasting immunity than do those inactivated by formalin or other means. The 17D attenuated yellow-fever virus gives immunity lasting for many years; it may even be life-long. On the other hand smallpox vaccination, even though it is a living vaccine, has to be repeated every few years to make sure of lasting immunity. Vaccination with either attenuated or killed vaccines may be followed by permanent immunity, for the subject may encounter virus in the normal course of life while he still has considerable immunity, and this contact can act as a booster and turn transient or partial immunity into something solid and permanent.

Every virus, however, presents its special problem. Vaccination against influenza (see p. 58) is rendered difficult by the antigenic changes which the virus undergoes. The multiplicity of serological types of rhinoviruses greatly reduces our hope of preventing colds by means of inoculation. Vaccination may give benefit at second-hand as it were: rabies in man will be prevented, not so much by immunizing people as by vaccinating the dogs which might bite them. Vaccination of British cattle against foot and mouth disease is, for reasons we have discussed (see p. 122), not the best solution. It is of more value to vaccinate the cattle in Europe or South America which are potential sources of introduction of virus into Britain.

On the whole attempts at specific cure of virus-diseases have had little success. Most virus infections have got a good grip on their victim before the disease is discovered and the virus is then inside cells and therefore difficult to destroy without destroying the body's cells also. For this reason sero-therapy, cure by administering antiserum, is of virtually no practical use. Giving of interferon, though an interesting theoretical possibility, has not yet given definite evidence that it will work in practice. Chemotherapy, so marvellously successful against many protozoal and bacterial infections, has a much harder task if it has to tackle a virus entrenched in its intracellular fastnesses. Some slight success has been achieved against infections of the cornea of the eye by herpes and vaccinia viruses [93]. The drugs used have been modifications of uridine, one of the bases of which nucleic acid is composed. The most notable achievement has been with a substance known as N-methyl-isatin-thiosemicarbazone; this when given to contacts of small-pox patients in India apparently reduced the incidence of the disease [14]. Only three mild cases appeared among 1,151 close contacts treated with the drug, while seventy-eight cases with twelve deaths occurred in a comparable control group. This, be it noted, was not chemotherapy but chemoprophy-laxis, the drug being given before the disease was expected to develop. Of other drugs the most that can be said is that they have given mildly encouraging results in laboratory tests, but have yet to prove their worth in the field.

It will be apparent from what has been written in this and preceding chapters that possibilities of controlling virus-diseases depend, above all, on understanding their ecology. This is often a complex matter involving co-operation by people trained in different disciplines. The thing to remember is that viruses, like other creatures, have to struggle for existence: their survival is by no means assured but must depend on a number of favourable circumstances. Often there is a weak link in the chain of causation of a virus-disease: it is up to virologists and their colleagues to discover it and act accordingly. Some viruses which have achieved a quiet yet unstable state of latency within their host will be very difficult to attack. Yet others have been successfully conquered and with more understanding of virus ecology, further conquests will come.

Appendix 1

There is at present considerable activity on an international basis in an effort to find a rational method of classifying all viruses. The three main criteria are: nature of nucleic acid (DNA or RNA), symmetry (cubical, helical or binary – i.e. a combination of the two) and presence or absence of envelope. The following tentative scheme has been put forward for discussion by an international committee.

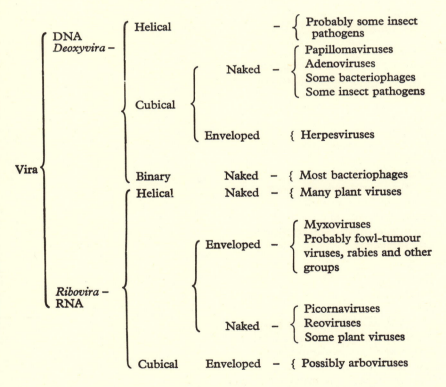

Vira

DNA
Deoxyvira –

- Helical – { Probably some insect pathogens
- Cubical
  - Naked – { Papillomaviruses / Adenoviruses / Some bacteriophages / Some insect pathogens
  - Enveloped { Herpesviruses
- Binary — Naked – { Most bacteriophages

Ribovira –
RNA

- Helical
  - Naked – { Many plant viruses
  - Enveloped – { Myxoviruses / Probably fowl-tumour viruses, rabies and other groups
  - Naked – { Picornaviruses / Reoviruses / Some plant viruses
- Cubical — Enveloped – { Possibly arboviruses

Poxviruses have not been placed as the nature of their symmetry is still uncertain.

207

Appendix 2

The chief families of viruses pathogenic for vertebrates are listed below together with some of their more important members, most of which have been mentioned in this book.

DNA viruses

Papovaviruses (Papillomaviruses)
    Wart-viruses of man, rabbit and other species
    Polyoma
    Vacuolating virus (SV 40)

Adenoviruses
    Thirty-one serotypes infecting man
    Others infecting monkeys, pigs, mice and fowls
    Infectious hepatitis of dogs

Herpesviruses
    Herpes simplex
    Varicella and herpes zoster
    B virus of monkeys
    Pseudorabies
    Infectious bovine rhinotracheitis
    Equine rhinopneumonitis
    Malignant catarrh of cattle
    Cytomegaloviruses (salivary viruses)
    Frog carcinoma

Poxviruses
    Smallpox
    Vaccinia
    Cowpox
    Infectious ectromelia (mouse-pox)
    Swine-, sheep-, goat-, horse-, camel-pox
    Orf (contagious pustular dermatitis)
    Fowl-pox
    Myxomatosis and fibromatosis of rabbits
    Yaba virus
    Molluscum contagiosum

RNA viruses

Picornaviruses
Poliomyelitis
Echoviruses ⎫
Coxsackieviruses ⎭ Human enteroviruses
Animal enteroviruses
Rhinoviruses
Foot-and-mouth disease
Encephalomyocarditis

Reoviruses
Three serological types infecting numerous species

Arboviruses
A group   Equine encephalomyelitis (Eastern, Western and Venezuelan)
Chikungunya
O'nyong-nyong

B group   Yellow fever
Japanese B
St Louis
West Nile
Louping-ill and tick-borne encephalitis
Kyasanur forest disease

Other groups Phlebotomus fever
Colorado tick fever
Rift valley fever

Myxoviruses
Influenza A, B and C
Para-influenza, 1, 2, 3 and 4
Newcastle disease
Mumps
Measles
Dog distemper
Rinderpest (cattle-plague)
Respiratory syncytial virus

Other RNA viruses which may come to be classified near the myxoviruses
Rabies
Fowl tumours and leucosis
Mammary tumour virus (Bittner)
Mouse leukaemia
Lymphocytic choriomeningitis
Rubella
Infectious bronchitis of fowls
Vesicular stomatitis
Visna

Viruses mentioned in this book but not yet classified
  Infectious and serum hepatitis
  African horse-sickness
  Blue-tongue
  Equine anaemia
  Scrapie

Glossary*

adenovirus: one of a family of deoxyviruses having affinity for glandular tissue.

alatae: winged forms of aphids.

allantois (allantoic): embryonic sac into which waste-products are excreted.

amnion (amniotic): sac immediately surrounding the embryo.

antibody: globulins (proteins of high molecular weight) stimulated by and reacting with antigens.

antigen: chemical substances, usually protein, stimulating formation of and reacting with antibody.

antigenic drift: gradual change to be found in antigenic make-up of certain viruses.

antigenic types: strains of virus differing from each other in antigenic make-up.

apterae: wingless forms of aphids.

arachnids: a family of arthropods containing ticks, spiders and mites.

arbovirus (arborvirus): a telescoped form of arthropod-borne virus.

arthropod: invertebrates with jointed limbs; a group containing insects, arachnids and crustaceans.

attenuation: weakening of the virulence of a virus.

bacteriophages: viruses which infect bacteria.

blind passage: serial passage of an agent 'by faith', with no visible evidence that it is multiplying.

capsid: the geometrically organized outer shell of a virus with cubical symmetry.

capsomeres: the units making up a capsid.

carcinogenic: causing cancer.

chorio-allantoic membrane: a membrane surrounding the allantoic cavity of an avian embryo.

cicadellids: members of a family of plant-bugs, the leaf-hoppers.

complement fixation: fixation by an antigen-antibody complex of a substance present in fresh serum.

conjunctivitis: inflammation of the superficial membrane covering the eye.

cytomegaloviruses: members of a group of herpesviruses, which cause infected cells to enlarge.

cytopathic (cytopathogenic): causing damage to or destruction of cells, especially in tissue culture.

deoxyvirus: virus of which the nucleic acid is of the de-oxy-type (DNA).

echoviruses: enteric cytopathogenic human, orphan viruses: members of the picornavirus family.

ecology: study of organisms in relation to their environment.

encephalitis: inflammation of the brain.

---

* The definitions are valid only in the context of their use in this book. Some may have wider connotations elsewhere.

enteroviruses: picornaviruses which infect the alimentary canal: those infecting man are poliomyelitis, coxsackie and echoviruses.

epizootic: an epidemic affecting animals other than man.

erythroblastosis: malignant disease involving the tissues forming red blood cells.

extrinsic incubation period: the period between acquisition of a parasite by an arthropod and beginning of transmissibility.

haemadsorption: the adsorption of red blood cells on to the surface of cells infected by certain viruses.

haemagglutination: clumping of red blood cells, particularly in the presence of certain viruses or virus products.

HeLa cells: a line of cells derived from a human cancer.

hepatitis: inflammation of the liver.

herpesviruses: a family of deoxyviruses related to that of herpes simplex.

icosahedron: a twenty-sided figure.

inclusions: bodies to be found in nucleus or cytoplasm of cells infected with certain viruses; they may or may not represent intracellular colonies of virus.

interferon: a low-molecular-weight protein formed by cells in response to the stimulus of virus, particularly dead or damaged virus.

intrinsic incubation period: period between infection of an animal or plant and the appearance of symptoms.

leukaemia (leucosis): malignant disease involving tissues which form white blood cells (leucocytes).

link host: a host, infection of which forms a br..lge between infection in a maintenance or reservoir host and one, such as man, infected only incidentally.

lymphoma, lymphomatosis, lymphosarcoma: malignant disease involving lymphatic tissue.

lysogeny: the state in which a bacteriophage is latent, integrated with the genetic apparatus of a bacterium, yet capable of being activated by certain stimuli.

maintenance host: one which in nature is concerned in the permanent existence of a parasite.

mycoplasma: a small, often filterable, bacterium without a conventional cell wall: also called pleuro-pneumonia-like organism or PPLO.

myeloblastosis: malignant disease involving precursors of granular leucocytes or polymorphs.

myxoviruses: members of a family of riboviruses having a helical symmetry in their nucleoprotein; those originally included had an affinity for mucins on the surface of red blood and other cells.

nematode: eelworm.

neurolymphomatosis: a disease of birds, with, as a rule, thickening of nerves due to dense infiltration with lymphocytes: otherwise called fowl paralysis or Marek's disease.

neurotropic: having an affinity for the nervous system.

nucleocapsid: the complex of nucleoprotein and its surrounding capsid.

oncogenic: causing tumours.

orphan: a name given to viruses which have no known association with disease.

ovipositor: the egg-laying organ of an insect.

papillomaviruses: deoxyviruses causing warts.

papovavirus: a name applied to the family of viruses causing papillomas and some other conditions:

passengers: viruses being carried along in tumours, to which they have no causal relation.

pathogen (pathogenic): causing disease.

phage: a short form of bacteriophage.

phloem: one of the conducting systems in plant-tissues.

picornaviruses: a family of very small riboviruses.

polyhedra: crystalline bodies occurring in many virus infections of insects; virus-particles are contained within them.

polyoma: a condition in which tumours of various kinds are produced by the action of a particular deoxyvirus.

prophage: bacteriophage in a state of integration with the genetic apparatus of a bacterium.

protozoa: unicellular animals.

encephalo-myelitis: inflammation of the brain and spinal cord.

reoviruses: members of a family of small riboviruses, having two-stranded RNA in their make-up.

rhinoviruses: acid sensitive picornaviruses, causing common colds.

ribonuclease: an enzyme splitting ribonucleic acid.

ribovirus: a virus of which the nucleic acid is of the ribose type (RNA).

rickettsia: a member of a family of small parasitic bacteria, usually intracellular and arthropod transmitted.

sarcoma: malignant tumour arising from connective tissues.

sentinels: animals tethered or caged in the open to see whether they will become naturally infected, as by the bite of mosquitoes.

serological types: same as antigenic types.

stylets: the piercing and sucking apparatus of aphids and other insects.

symbiosis: living together of two organisms with benefit to both.

syncytium (syncytial): running together of cells, resulting in production of very large cells with many nuclei.

taiga: the marshy forest zone of the far north.

target organ: organ to which a virus is localized or which it particularly attacks.

taxonomists: people concerned with the classification or orderly arrangement of organisms.

titre: highest effective dilution.

transformation: change in cells in culture, causing them to have some or all of the properties of malignant cells.

transovarial: passage through the egg to the next generation.

ungulates: hoofed animals.

vacuolating: producing vacuoles or cavities within cells.

vertical transmission: transmission of an infection from one generation to the next.

vector: agent carrying infection, usually applied to an arthropod causing infection of a vertebrate or plant.

viraemia: presence of virus in circulating blood.

zoonoses: infections transmitted from other animals to man.

Bibliography

The reader who wishes to learn more about particular viruses is advised to consult one or other of the following textbooks.

(i) F. L. HORSFALL & I. TAMM [Ed.] *Viral and rickettsial diseases of man* 4th ed. (Lippincott: Philadelphia)
This is the standard textbook on the viruses infecting man, and gives much information about symptoms and treatment as well as dealing with viral properties and ecology.

(ii) S. P. BEDSON, A. W. DOWNIE, F. O. MacCALLUM & C. H. STUART HARRIS (1961) *Virus and rickettsial diseases* 3rd ed. (Arnold: London)
This covers much the same ground as (i) but much more consisely.

(iii) A. J. RHODES & C. E. VAN ROOYEN (1958) *Text-book on virology* (4th ed. in preparation) (Williams & Wilkins: Baltimore)
This again covers similar ground and is intermediate in size and cost between (i) and (ii).

There is much less information readily available about viruses affecting animals other than man. There are, however, the following books.

(iv) C. H. ANDREWES (1964) *Viruses of vertebrates* (Baillière, Tindall & Cox: London)   (Second edition due to be published in 1967)
This gives very concisely the more important facts concerning the properties of all known viruses infecting vertebrates, together with some information about pathology and ecology.

(v) A. O. BETTS & C. J. YORK [Ed.] *Viral and rickettsial diseases of animals.* To be published shortly in two volumes (Academic Press: New York)

There are chapters on viruses in:

(vi) W. A. HAGAN & D. W. BRUNER (1961) *Infectious diseases of domestic animals* 4th ed. (Baillière, Tindall & Cox: London)

In the field of plant viruses we have:

(vii) F. C. BAWDEN (1964) *Plant viruses and virus diseases* 4th ed. (Ronald: New York)

(viii) M. K. CORBETT & H. D. LISTER (1964) *Plant virology* (Univ. of Florida Press, Gainville)
This is a fairly detailed textbook.

(ix) K. M. SMITH (1957) *Textbook of plant and virus diseases* (Churchill: London)
Essentially a descriptive catalogue with pictures.

Books of a more general nature:

(x) F. M. BURNET (1960) *Principles of animal virology* 2nd ed. (Academic Books: London)

Deals, as the title indicates, with general aspects of viral disease rather than with viruses individually.

(xi) F. M. BURNET & W. M. STANLEY [Ed.] (1959) *The viruses* Three vols. (Academic Press: New York & London)

An advanced textbook on general virology dealing with fundamental principles (Vol. I), with particular aspects of plant viruses and phages (Vol. II) and animal viruses (Vol. III).

(xii) A. P. WATERSON (1961) *Introduction to animal virology* (Cambridge University Press)

A simple and concise introduction to general principles, particularly at the cellular level.

(xiii) *British Medical Bulletin* (British Council: London)

Summaries of current knowledge about most of the important virus diseases of man are to be found in Vol. 9, no. 3 (1953) and Vol. 15, no. 3 (1959). Another virus number is due for publication in May 1967.

References

1 AHLSTRÖM, C. G. & FORSBY, N. (1962) Sarcomas in hamsters after injection with Rous chicken tumour material. *J. exp. Med.* **115**, 839–852

2 AHLSTRÖM, C. G., & JONSSON, N. (1962) Induction of sarcoma in rats by a variant of Rous virus. *Acta path. microbiolog. scand.* **54**, 145–172

3 ALMEIDA, J.D., HOWATSON, A.F., PINTERIC, L., & FENJE, P. (1962) Electron microscope observations on rabies virus by negative staining. *Virology* **18**, 147–151

4 ANDREWES, C.H. (1950) Adventures among viruses III. The puzzle of the common cold. *New Engl. J. Med.* **242**, 235–240

5 ANDREWES, C.H. (1952) The work of the World Influenza Centre. *J. Roy. Inst. publ. Health* **15**, 309–318

6 ANDREWES, C.H. (1954) The immunological problem of influenza. *Practitioner* **173**, 534–539

7 ANDREWES, C.H. (1957) Factors in virus evolution. *Adv. in Virus Research* **4**, 1–24

8 ANDREWES, C.H. (1959) Asian influenza: a challenge to epidemiology; in *Perspectives in virology* 184–191 (Chapman & Hall: London)

9 ANDREWES, C.H. (1964) The complex epidemiology of respiratory virus infections. *Science* **146**, 1274–1277

10 ANDREWES, C.H. & CARMICHAEL, E.A. (1930) A note on the presence of antibodies to herpes virus in post-encephalitic and other human sera. *Lancet* **1**, 857–858

11 ANDREWES, C.H. & MILLER, C.P. (1924) A filterable virus infection of rabbits. II. Its occurrence in apparently normal rabbits. *J. exp. Med.* **40**, 789–796

12 ANDREWES, C.H., THOMPSON, H.V. & MANSI, W. (1959) Myxomatosis: present position and future prospects in Great Britain. *Nature* **184**, 1179–1180

13 BARON, S. & ISAACS, A. (1961) Mechanism of recovery from viral infection in the chick embryo. *Nature* **191**, 97–98

14 BAUER, D.J., ST VINCENT, L., KEMPE, C.H. & DOWNIE, A.W. (1963) Prophylactic treatment of smallpox contacts with N-methylisatin-β-thiosemicarbazone. *Lancet* **2**, 494–496

15 BAWDEN, F. (1950) *Plant viruses and virus diseases.* 3rd ed. (Chronica Botanica: Waltham)

16 BAWDEN, F., KASSANIS, B. & NIXON, H.L. (1950) The mechanical trans-

mission and some properties of potato paracrinkle virus. *J. gen. Microbiol.* **4,** 210–219

17 BECKER, W., NAUDÉ, W. duT. & KIPPS, A. (1963) Virus studies in disseminated herpes simplex infections. *S. African med. J.* **37,** 74–76

18 BERNKOPF, H., LEVINE, S. & NERSON, R. (1953) Isolation of West Nile Virus in Israel. *J. inf. Dis.,* **93,** 207–218

19 BIGGS, P.M. (1961) A discussion on the classification of the avian leucosis complex and fowl paralysis. *Brit. vet. J.* **117,** 326–334

20 BIGGS, P.M. & PAYNE, L.N. (1963) Transmission experiments with Marek's disease (fowl paralysis). *Vet. Rec.* **75,** 177–179

21 BITTNER, J.J. (1948) Some enigmas associated with genesis of mammary cancer in mice. *Cancer Research* **8,** 625–639

22 BLACK, L.M. (1953) Transmission of plant viruses by cicadellids. *Adv. Virus Research* **1,** 68–89

23 BOYD, J.S.K. (1959) Yellow fever and its prevention. *J. Royal Inst. publ. Health & Hyg.* **22,** 11–26

24 BRADISH, C.J. & KIRKHAM, J.B. (1966) *J. gen Microbiol.* **44,** 359–371

25 BREITENFELD, P.M. & SCHÄFER, W. (1957) The formation of fowl plague virus antigen in infected cells as studied with fluorescent antibodies. *Virology* **4,** 328–345

26 BROADBENT, L. & MARTINI, C. (1959) The spread of plant viruses. *Adv. Virus Research* **6,** 93–135

27 BUCKLAND, F.E. & TYRRELL, D.A.J. (1962) Loss of infectivity on drying various viruses. *Nature* **195,** 1063–1064

28 BUCKLAND, F.E. & TYRRELL, D.A.J. (1964) Experiments on the spread of colds. I. Laboratory studies on the dispersal of nasal secretion. *J. Hyg.* **62,** 365–377; also (1965) II. Studies in volunteers with Coxsackie A21. *J. Hyg.* **63,** 327–343

29 BURKITT, D. (1958) A sarcoma involving the jaws of African children. *British J. Surgery* **46,** 218–223

30 BURMESTER, B.R. (1957) Transmission of tumour-inducing avian viruses under natural conditions. *Texas reports Biol. & Med.* **15,** 540–558

31 BURNET, F.M. (1955) *Principles of animal virology* 445 (Academic Press: New York)

32 *Ibid.* 267 and 451

33 *Ibid.* 409

34 *Ibid.* 385

35 BURNET, F.M. & WILLIAMS, S.W. (1939) Herpes simplex: a new point of view. *Med. J. Australia* **1,** 637–642

36 CARTER, W. (1961) Ecological aspects of plant virus transmissions. *Ann. rev. Entomology* **6,** 347–370

37 CASALS, J. (1964) Antigenic variants of eastern equine encephalitis virus. *J. Exp. Med.* **119,** 547–565

38 CAUSEY, O.R. & SHOPE, R.E. (1965) Icoaraci, a new virus related to Naples phlebotomus fever virus. *Proc. Soc. exp. Biol. (NY)* **118**, 420–421

39 CHAPPLE, P.J. & LEWIS, N.D. (1965) Myxomatosis and the rabbit flea. *Nature* **207**, 388–389

40 CHAPRONIERE, D.M. & ANDREWES, C.H. (1957) Cultivation of rabbit myxoma and fibroma in tissues of non-susceptible hosts. *Virology* **4**, 351–365

41 CLARKE, D.H. (1962) in *Symposium on the biology of Viruses of the tick-borne encephalitis complex.* 67 (Academic Press: New York); also (1964) Further studies on antigenic relationships among the viruses of the group B tick-borne complex. *Bull. Wld. Hlth. Org.* **31**, 45–56

42 CLELAND, J.W. & CAMPBELL, A.W. (1919–20). An experimental investigation of an Australian epidemic of acute encephalomyelitis. *J. Hyg.* **18**, 272–316

43 COLLIER, L.H. (1959) Recent advances in the virology of trachoma and inclusion conjunctivitis and allied diseases. *Brit. med. Bull.* **15**, 231–234

44 COONS, A.H. & KAPLAN, M.H. (1950) Localization of antigen in tissue cells. II. Improvements in a method for the detection of antigen by means of fluorescent antibody. *J. exp. Med.* **81**, 1–29

45 DE HARVEN, E. & FRIEND, C. (1964) Structure of virus particles partially purified from the blood of leukaemic mice. *Virology* **23**, 119–124

46 DERRICK, E.H. (1964) The epidemiology of Q-fever. *J. Hyg.* **43**, 357–361

47 DERRICK, E.H. & BICKS, V.A. (1958) Limiting temperature for the transmission of dengue. *Aust. ann. med.,* **7**, 102–107

48 DE TRAY, D.E. (1963) African swine fever. *Adv. vet. Science.* **8**, 299–333

49 DODD, K., JOHNSTON, L.M. & BUDDINGH, G.J. (1938) Herpetic stomatitis. *J. Pediatrics* **12**, 95–102

50 DOWLING, H.F., JACKSON, G.G., SPIESMAN, I.G. & INOUYE, T. (1958) Transmission of the common cold to volunteers under controlled conditions. III. The effect of chilling of the subjects upon susceptibility. *Amer. J. Hyg.* **66**, 59–65

51 DUDLEY, S.F. (1934) Can yellow fever spread into Asia ? *J. trop. Med. & Hyg.* **37**, 273–278

52 EDDY, B. (1960) The polyoma virus. *Adv. Virus Research* **7**, 91–102

53 EICHENWALD, H.F., KOTSEVALOV, O. & FASSO, L.A. (1960) The 'cloud-baby': an example of bacterial-viral interaction. *Amer. J. Dis. Children* **100**, 161–173

54 ENRIGHT, J.B. (1956) Bats and their relation to rabies. *Ann. Rev. Microbiol.* **10**, 369–392

55 ENDERS, J.F., WELLER, T.H. & ROBBINS, F.C. (1949) Cultivation of the Lansing strain of poliomyelitis in cultures of various human embryonic tissues. *Science* **109**, 85–87

56 EPSTEIN, M.A., ACHONG, B.G. & BARR, Y.M. (1964) Virus particles from cultured lymphocytes from Burkitt's lymphoma. *Lancet* **1**, 702–703

57 FENNER, F. (1948) The pathogenesis of the acute exanthems. *Lancet* **2**, 915–920

58 FENNER, F. & RATCLIFF, F.N. (1965) *Myxomatosis.* (Camb. Univ. Press.)

59 FENNER, F. & CHAPPLE, P.J. (1965) Evolutionary changes in myxoma virus in Britain. *J. Hyg.* **63**, 175–185

60 GAJDUSEK, D. C. (1962) Virus haemorrhagic fevers. *J. Pediatrics* **60**, 841–857

61 GEAR, J. H. S. (1958) Coxsackie virus infections of the newborn. *Progr. med. Virology* **1**, 106–121

62 GLEDHILL, A. W. (1962) Latent ectromelia. *Nature* **196**, 298

63 GOLDFIELD, M. & SUSSMAN, O. quoted by Shope in *Viral & Rickettsial Diseases of Man* 4th Ed. 391. Ed. Horsfall & Tamm (Lippincott: Philadelphia)

64 GOMATOS, P. J. & TAMM, I. (1963) The secondary structure of reovirus RNA. *Proc. Nat. Acad. Sci. (Wash.)* **49**, 707–714

65 GORDON, W. S., BROWNLEE A., WILSON, D. R. & MACLEOD, J. (1962) The epizootiology of louping-ill and tick-borne fever with observations on the control of these sheep-diseases. *Symp. Zool. soc. London* No. 6, 1–27

66 GREEN, R. G. (1935) on the nature of filterable viruses. *Science* **82**, 443–445

67 GREGG, N. M. (1941) Congenital cataract following German measles in the mother. *Trans. ophthalm. Soc. Australia*, **3**, 35–46

68 GROSS, L. (1951) Pathogenic properties and 'vertical' transmission of the mouse leukaemia agent. *Proc. Soc. exp. Biol. (NY)* **78**, 342–348

69 GROSS, L. (1963) Properties of a virus isolated from leukaemic mice, including various forms of leukaemia and lymphomas in mice and rats. *Viruses, nucleic acids and cancer* 403–426 (Williams, Wilkins: Baltimore)

70 HABEL, K. (1961) Resistance of polyoma-virus immune animals to transplanted polyoma tumours. *Proc. Soc. exp. Biol. (NY)* **106**, 722–725

71 HAMMON, W. McD., RUDNICK, A. & SATHER, G. E. (1960) Viruses associated with haemorrhagic fevers of the Philippines and Thailand. *Science* **131**, 1102–1103

72 HANAFUSA, H., HANAFUSA, T. & RUBIN, H. (1963) The defectiveness of the Rous sarcoma virus. *Proc. Nat. Acad. Sci. (Wash.)* **49**, 572–580

73 HARRISON, B. D., (1960) The biology of the soil-borne plant viruses. *Adv Virus Research* **7**, 131–161

74 HARRISON, B. D. & WINSLOW, R. D. (1961) Laboratory and field studies on the relation of arabis mosaic virus to its nematode vector *Xiphinema diversicaudatum*, Micoletzky. *Ann. appl. Biol.* **49**, 621–633

75 HAYES, R. O., BEADLE, L. D., HESS, A. D., SUSSMAN, O. & BONESE, M. J. (1962) Entomological aspects of the eastern equine encephalomyelitis New Jersey outbreak. *J. trop. Med. & Hyg.* **11**, 115–121

76 HEMMES, J. H., WINKLER, K. C. & KOOL, S. M. (1960) Virus survival as a seasonal factor in influenza and poliomyelitis. *Nature* **188**, 430–431

77 HERSHEY, A. D. (1957) Bacteriophages as genetic and biochemical systems. *Adv. Virus Research* **4**, 25–61

78 HIRST, G. K. (1941) The agglutination of red cells by allantoic fluid of chick embryos infected with influenza virus. *Science* **94**, 22–23

79 HOPE-SIMPSON, R. E. (1958) Discussion on the common cold. *Proc. Roy. Soc. Med.* **51**, 267–272

80 HOPE-SIMPSON, R. E. (1965) The nature of herpes zoster: a long-term study and a new hypothesis. *Proc. Roy. Soc. Med.* **58**, 9–20

81 HORNE, R. W. & WILDY, P. (1962) Recent studies on the fine structure of

viruses by electron-microscopy, using negative staining techniques. *Brit. med. Bull.* **18**, 199–204

82 HUEBNER, R. J. (1947) Report of an outbreak of Q-fever at National Institute of Health. II. Epidemiological features. *Amer. J. publ. Health* **37**, 431–440

83 HUEBNER, R. J., JELLISON, W. L. & BECK, M. D. (1949) Q-fever: a review of current knowledge. *Ann. int. Med.* **30**, 495–509

84 HUEBNER, R. J., ROWE, W. P., HARTLEY, J. W. & LANE, W. T. (1962) Mouse polyoma in a rural ecology. *Tumour viruses of murine origin.* (Churchill: London)

85 HULL, R. N., MINNER, J. R. & MASCOLI, C. C. (1956) New viral agents recovered from monkey tissue culture cells. I. *Amer. J. Hyg.* **63**, 204–215; also III. (1958) *Amer. J. Hyg.* **68**, 31–44

86 HURST, E. W. & PAWAN, J. L. (1931) An outbreak of rabies in Trinidad without history of bites and with symptoms of acute ascending myelitis. *Lancet* **2**, 622–628

87 ISAACS, A. (1961) Interferon. *Scientific American* **204**, 51–57

88 JACKSON, G. G. & DOWLING, H. F. (1959) Transmission of the common cold to volunteers under controlled conditions. IV. Specific immunity to the common cold. *J. clin. invest.* **38**, 762–769

89 JOHNSON, H. N. (1965) *Rabies: in virus and Rickettsial diseases of man.* 4th Ed. Ed. Horsfall & Tamm (Lippincott: Philadelphia)

90 JOHNSON, R. T. (1964) Pathogenesis of herpes virus encephalitis. II. A cellular basis for the development of resistance with age. *J. exp. Med.* **120**, 359–373

91 JONKERS, A. H., SPENCE, C. A., COAKWELL, C. A. & THORNTON, J. J. (1964) Laboratory studies with wild rodents and viruses native to Trinidad. *Amer. J. trop. Med. & Hyg.* **13**, 613–619

92 KASSANIS, B. & MACFARLANE, I. (1964) Transmission of tobacco necrosis virus by zoospores of *Olpidium brassicae*. *J. gen. Microbiol.* **36**, 79–93

93 KAUFMAN, H. E. (1965) Problems in virus chemotherapy. *Progr. med. Virology* **7**, 116–159

94 KIDD, J. G. & ROUS, P. (1938) The carcinogenic effect of a papilloma virus on the tarred skin of rabbits. II. *J. exp. Med.* **68**, 529–562

95 KILHAM, L., HERMAN, C. M. & FISHER, E. R. (1953) Naturally occurring fibromas of grey squirrels related to Shope's rabbit fibroma. *Proc. Soc. Exp. Biol.* (*NY*) **82**, 298–301

96 LAIDLAW, P. P. (1938) *Virus diseases and viruses* (Camb. Univ. Press)

97 LEHANE, O., KWANTES, C. M. S., UPWARD, M. G. & THOMPSON, D. R. (1949) Homologous serum jaundice. *Brit. med. J.* **2**, 572–754

98 LIDWELL, O. M. & SOMMERVILLE, T. (1951) Observations on the incidence and distribution of the common cold in a rural community during 1948 and 1949. *J. Hyg.* **49**, 365–381

99 LIDWELL, O. M. & WILLIAMS, R. E. O. (1961) The epidemiology of the common cold. *J. Hyg.* **59**, 309–319 and 321–334

100 LOVELOCK, J. E., PORTERFIELD, J. S., RODEN, A. T., SOMMERVILLE, T. & ANDREWES, C. H. (1952) Further studies on the natural transmission of the common cold. *Lancet* **2**, 657–660

101 LUCKÉ, B. (1938) Carcinoma in the leopard frog: its probable causation by a virus. *J. exp. Med.* **68**, 457–468

102 McCARTHY, K., TAYLOR-ROBINSON, C.H. & PILLINGER, S.E. (1963) Isolation of rubella virus from cases in Britain. *Lancet* **2**, 593–598

103 MACKENZIE, R.B., BEYE, H.K., VALVERDA, L., & GARRON, H. (1964) Epidemic haemorrhagic fever in Bolivia. *Amer J. trop. Med. Hyg.* **13**, 620–625

104 McLAUCHLAN, J.D. & HENDERSON, W.M. (1947) The occurrence of foot-and-mouth disease in the hedgehog under natural conditions. *J. Hyg.* **45**, 474

105 MAGRASSI, F. quoted by CHU C.M., ANDREWES, C.H. & GLEDHILL, A.W. (1950) *Bull. World Health Org.* **3**, 127–214

106 MAITLAND, H.B. & MAITLAND, M.C. (1928) Cultivation of vaccinia virus without tissue culture. *Lancet*, **2**, 596–597

107 MARAMOROSCH, K. (1952) Direct evidence for the multiplication of aster yellows virus in its insect vector. *Phytopathology* **42**, 59–64

108 MARAMOROSCH, K. (1963) Arthropod-transmission of plant viruses. *Ann. Rev. Entomology* **8**, 369–414

109 MARAMOROSCH, K. & JENSEN, D.D. (1963) Harmful and beneficial effects of plant viruses in insects. *Ann. Rev. Microbiology* **17**, 495–530

110 MARSHALL, I.D. & FENNER, F. (1960) Studies in the epidemiology of infectious myxomatosis of rabbits IV. *J. Hyg.* **58**, 485–488

111 MARSHALL, I.D. & DOUGLAS, G.W. (1961) *idem* VIII. *J. Hyg.* **59**, 117–122

112 MARSHALL, I.D. & REGNERY, D.C. (1963) Studies on the epidemiology of myxomatosis in California III. *Amer. J. Hyg.* **77**, 213–319

113 MEAD-BRIGGS, A.R. (1964) Some experiments concerning the interchange of rabbit fleas, *Spilopsyllus cuniculi*, (Dale) between living rabbit hosts. *J. animal Ecol.* **33**, 13–26

114 METTLER, N., BUCKLEY, S.J. & CASALS, J. (1961) Propagation of Junin virus, the etiological agent of Argentinian haemorrhagic fever in HeLa cell cultures. *Proc. Soc. exp. Biol.* (*NY*) **107**, 684–688

115 MEYER, K.F. & EDDIE, B. (1958) Ecology of avian psittacosis especially in parakeets. *Progress in psittacosis research and control* 52–79 (Rutgers Univ. Press: New Brunswick)

116 MILAM, D.F. & SMILLIE, W.G. (1931) A bacteriological study of 'colds' on an isolated tropical island (St John). *J. exp. Med.* **53**, 733–752

117 MILES, J.A.R. (1964) Some ecological aspects of the problem of arthropod-borne animal diseases in the Western Pacific and South-East Asia regions. *Bull. World. Health Org.* **30**, 197–210

118 MULDER, J. & MASUREL, N. (1958) The presence of pre-epidemic antibody against the 1957 strain of Asiatic influenza in the serum of older people living in the Netherlands. *Lancet* **1**, 810–814

119 MYKYTOWYCZ, R. (1953) An attenuated strain of the myxomatosis virus recovered from the field. *Nature* **172**, 448–449

120 NIVEN, J.S.F., ARMSTRONG, J.A., ANDREWES, C.H., PEREIRA, H.G. & VALENTINE, R.C. (1961) Subcutaneous 'growths' in a monkey produced by a poxvirus. *J. Path. Bact.* **81**, 1–14

121 OLIN, G. (1952) The epidemiologic pattern of poliomyelitis in Sweden from

1905–1950. *Poliomyelitis: papers and discussions presented at the 2nd International Poliomyelitis Congress* 367–375 (Lippincott: Philadelphia)

122 PARKER, F. & NYE, R.N. (1925) Studies on filterable viruses I. Cultivation of vaccinia virus. *Amer. J. Path.* **1**, 325–335

123 PARKMAN, P.D., BUESCHER, E.L. & ARTENSTEIN, M.S. (1962) Recovery of rubella virus from army recruits. *Proc. Soc. exp. Biol.* (*NY*) **111**, 225–230

124 PAUL, J.H. & FREESE, H.L. (1933) Epidemiological and bacteriological study of 'common cold' in isolated Arctic community (Spitzbergen). *Amer. J. Hyg.* **17**, 517–535

125 PAUL, J.R., MELNICK, J.L., BENNETT, V.H. & GOLDBLUM, N. (1952) Survey of neutralizing antibodies to poliomyelitis in Cairo, Egypt. *Amer. J. Hyg.* **55**, 402–413

126 PETERS, J.T. (1954) Equine infectious anaemia transmitted to man. *Ann. int. Med.* **23**, 271–274

127 PIERCE, W.E., STILLE, W.T. & MILLER, L.F. (1963) A preliminary report on effects of routine inoculations on respiratory illness. *Proc. Soc. exp. Biol.* (*NY*) **114**, 369–372

128 PIERCY, S.F. (1961) The menace of African horse sickness. *New Scientist* **9**, 76–78

129 PLOWRIGHT, W., FERRIS, R.D. & SCOTT, G.R. (1960) Blue wildebeest and the aetiological agent of bovine malignant catarrhal fever. *Nature* **188**, 1167–1169

130 Provisional Committee for Virus Nomenclature (1965) *Ann. Inst. Pasteur* **109**, 625–637

131 RAFFERTY, K.A. (1964) Kidney tumours of the leopard frog. *Cancer Research* **24**, 169–185

132 RAMOS-ALVAREZ, M. & SABIN, A.B. (1956) Intestinal viral flora of healthy children demonstrated by monkey kidney tissue culture *Amer. J. publ. Health* **46**, 295–299

133 REEVES, W.C., BELLAMY, R.E. & SCRIVANI, R.P. (1958) Relationship of mosquito vectors to winter survival of encephalitis virus. I. Under natural conditions. *Amer. J. Hyg.* **67**, 78–89

134 RIVERS, T.M. & TILLETT, W.S. (1923) Studies on varicella. *J. exp. Med.* **38**, 673–690; *ibid.* **39**, 777–802

135 ROGERS, S. & ROUS, P. (1951) Joint action of a chemical carcinogen and a neoplastic virus to induce cancer in rabbits. *J. exp. Med.* **93**, 459–485

136 ROSS, R.W. (1956) The Newala epidemic. III. The virus, isolation, pathogenic properties and relationship to the epidemic. *J. Hyg.* **54**, 177–191

137 ROTHSCHILD, M. (1965) Myxomatosis and the rabbit flea. *Nature* **207**, 1162–1163

138 ROWE, W.P., HARTLEY, J.W., WATERMAN, S., TURNER, H.E. & HUEBNER, R.J. (1958) Cytopathogenic agent resembling human salivary gland virus recovered from tissue-cultures of human adenoids. *Proc. Soc. exp. Biol.* (*NY*) **92**, 418–424

139 ROWE, W.P., HUEBNER, R.J., GILMORE, L.K., PARROTT, R.H. & WARD, T.G. (1953) Isolation of a cytopathogenic agent from human adenoids undergoing spontaneous degeneration in tissue-culture. *Proc. Soc. exp. Biol.* (*NY*) **81**, 570–573

140 RUBIN, H., FANSHIER, L., CORNELIUS, A. & HUGHES, W.F. (1962) Tolerance and immunity in chickens after congenital and contact infection with an avian leukosis virus. *Virology* **17**, 143–156

141 SCHERER, W.F. & BUESCHER, E.L. (1959) Ecological studies of Japanese encephalitis virus in Japan. *Amer. J. trop. Med. & Hyg.* **8**, 644–650 (and following papers in the same journal number)

142 SERIÉ, C., ANDRAL, L., LINDREA, A. & NERI, P. (1964) Epidemic of yellow fever in Ethiopia, 1960–62. *Bull. World Health Org.* **30**, 299–319

143 SEVERIN, H.H.P. (1946) Longevity and life-history of leaf hopper species on virus-infected and on healthy plants. *Hilgardia* **17**, 121–133

144 SHOPE, R.E. (1932) A transmissible tumour-like condition in rabbits. *J. exp. Med.* **56**, 793–802

145 SHOPE, R.E. (1935) Experiments on epidemiology of pseudorabies: mode of transmission of disease in swine and their possible role in its spread to cattle. *J. exp. Med.* **62**, 85–99

146 SHOPE, R.E. (1940) Swine-pox. *Arch. Virusforschung* **1**, 457–467

147 SHOPE, R.E. (1964) The epidemiology of the origin and perpetuation of a new disease. *Perspectives in Biol. & Med.* **7**, 263–278

148 SHORE, H. (1961) O'nyong-nyong fever: an epidemic virus disease in East Africa. *Trans. roy. Soc. trop. Med. Hyg.* **55**, 361–373

149 SIGURDSSON. (1954) Observations on three slow infections of sheep. *Brit. vet. J.* **110**, 255–270 and 341–354

150 SILBERSCHMIDT, K., FLORES, E. & TOMMASI, M.R. (1957) Further studies on the experimental transmission of 'infectious chlorosis' of Malvaceae. *Phytopathol.* **30**, 387–314

151 SMITH, C.E. GORDON (1956) The history of dengue in tropical Asia and its probable relationship to the mosquito *Aëdes aegypti*. *J. trop. Med. Hyg.* **59**, 243–251

152 SMITH, C.E. GORDON (1956) A virus resembling Russian spring-summer encephalitis from an Ixodid tick in Malaya. *Nature* **178**, 581–582

153 SMITH, C.E. GORDON (1962) Ticks and viruses. *Sympos. Zool. Soc. Lond.* no. 6, 199–221

154 SMITH, K.M. (1962) The arthropod viruses. *Adv. Virus Research* **9**, 195–240

155 SMITH, K.M. & RIVERS, C.F. (1956) Some viruses affecting insects of economic importance. *Parasitology* **46**, 235–242

156 SMITH, M.G. (1959) The salivary gland viruses of man and animals (Cytomegalic inclusion disease). *Progr. med. Virol.* **2**, 171–202

157 SMITH, W.S., ANDREWES, C.H. & LAIDLAW, P.P. (1933) A virus obtained from influenza patients. *Lancet* **2**, 66–68

158 SMITHBURN, K.C., HADDOW, A.J. & LUMSDEN, W.H.R. (1949) An outbreak of sylvan yellow fever in Uganda with *Aëdes africanus* as principal vector. *Ann. trop. Med. Parasitol.* **43**, 74–89

159 SOPER, F.L. (1938) Yellow fever: the present situation (October 1938) with special reference to South America. *Tr. Roy. Soc. trop. Med. Hyg.* **32**, 297–332

160 SOUTHAM, C.M. & MOORE, A.F. (1954) Induced virus infections in man by the Egypt isolates of West Nile virus. *Amer. J. trop. Med. Hyg.* **3**, 19–53

161 STOREY, H.H. (1933) Investigations of the mechanism of the transmission of plant viruses by insect vectors. *Proc. Roy. Soc.* **113B**, 463–485

162 STRODE, G.K. (1951) *Yellow fever* (McGraw-Hill: New York)

163 STUART-HARRIS, C.H. (1965) *Influenza and other virus infections of the respiratory tract* 2nd ed. (Arnold: London)

164 SWAN, C. (1964) Rubella in pregnancy as an etiological factor in congenital malformation, stillbirth, miscarriage and abortion. *J. Obst. & Gynae., Brit. Empire* **56**, 341–363 and 591–605

165 SWEET, H. & HILLEMAN, M.R. (1960) The vacuolating virus SV40. *Proc. Soc. exp. Biol. (N Y)* **105**, 420–427

166 THOMAS, A. S. (1960) Changes in vegetation since the advent of myxomatosis. *J. Ecol.* **48**, 287–306

167 THOMAS, L.A. & EKLUND, C.M. (1960) Overwintering of western equine encephalomyelitis virus in experimentally infected garter snakes and transmission to mosquitoes. *Proc. Soc. exp. Biol. (N Y)* **105**, 52–55

168 THORMAR, H. (1965) A comparison of visna and maedi viruses. *Res. in Vet. Science.* **6**, 117–129

169 TIERKEL, E. S. (1959) Rabies. *Adv. in Vet. Science* **5**, 183–226

170 TOBIN, J.O'H. (1963) Coxsackie viruses. *Brit. med. Bull.* **9**, 201–205

171 TRAUB, E. (1939) Epidemiology of lymphocytic choriomeningitis in mouse-stock observed for four years. *J. exp. Med.* **69**, 801–817

172 TRAUB, R., HERTIG, M., LAWRENCE, W.H. & HARRISS, T.T. (1954) Potential vectors and reservoirs of haemorrhagic fever in Korea. *Amer. J. Hyg.* **59**, 291–305

173 TYRRELL, D.A.J. & BYNOE, M.L. (1961) Some further virus isolations from common colds. *Brit. med. J.* **1**, 393–397

174 TYRRELL, D.A.J., BYNOE, M.L., HITCHCOCK, G., PEREIRA, H.G., ANDREWES, C.H. & PARSONS, R. (1960) Some virus isolations from common colds. *Lancet* **1**, 235–242

175 VAN LOGHEM, J.J. (1928) An epidemiological contribution to the knowledge of respiratory diseases. *J. Hyg.* **28**, 33–54

176 VAN TONGEREN, H.A.E. (1955) Encephalitis in Austria. IV. Excretion of virus by milk of the experimentally infected goat. *Arch. Virusforschung* **6**, 158–162

177 VOGEL, J.E. & SHELOKOV, A. (1957) Adsorption haemagglutination-test for influenza virus in monkey kidney tissue-culture. *Science* **126**, 358–359

178 WARNER, P. (1957) The detection of Murray valley encephalitis antibodies in hens' eggs. *Austral. J. exp. Biol. med. Sci.* **35**, 327–333

179 WATERSON, A.P. (1963) The origin and evolution of viruses. I. Virus origins: degenerate bacteria or vagrant genes. *New Scientist* **18**, 200–202

180 WELLER, T.H. (1965) Changing epidemiologic concepts of rubella with particular reference to unique characteristics of the congenital infection. *Yale J. Biol. & Med.* **37**, 455–467

181 WELLER, T.H. & HANSHAW, J.B. (1962) Virologic and chemical observations on cytomegalic inclusion disease. *New Engl. J. Med.* **266**, 1233–1244

182 WESTWOOD, J.C.N., ZWARTOUW, H.T., APPLEYARD, G. & TITMUSS, D.H.J. (1965) Comparison of the soluble antigens and virus particle antigens of vaccinia virus. *J. gen. Microbiol.* **38**, 47–53

183 WILDY, P. & HORNE, R.W. (1963) Structure of animal virus particles. *Progr. med. Virology* **5**, 1–12

184 WORK, T.H. (1958) Russian spring-summer virus in India: Kyasanur forest disease. *Progr. med. Virology* **1**, 248–279

185 YARWOOD, C.E. (1957) Mechanical transmission of plant viruses. *Adv. Virus research* **6**, 243–278

186 ZINSSER, H. (1935) *Rats, lice and history* (Routledge: London)

Index

Abutilon, 130, 133
Acidity, 41
Activation, 52, 163, 166, 174, 176
Aëdes aegypti, 86, 93–96, 99, 117, 142
Aëdes, other species, 77, 80, 85, 90, 93, 95–96, 120
Aerial transmissions, 48, 119
Africa
    Central, 85, 87, 98
    East, 85, 87, 106, 128, 147
    North, 149, 159
    South, 58, 60, 71, 76, 81, 87, 147, 164, 171
    West, 98, 106, 147, 192
Age and susceptibility, 69
Alabama, 82
Alatae, 126
Alberta, 112
Aleyrodidae, 130
Allantoic cavity (fluid), 20
Amazon, 77
Amnion, 20
Amplifiers, 89, 111
Andrewes, C.H., 10, 41, 163
Anopheles, 83, 85, 151–152, 156–158
Antelopes, 115, 121
Antibiotics, 3, 196
Antibodies, 34–38
Antigenic drift, 58–59
Antigenic types, 35, 57–61, 68, 116–117, 121
Ants, 28
Aphids, 27, 125–128, 131
Apterae, 126
Arachnids, 31, 128
Aragao H. de B., 151
Arctic, 46, 142
Argentina, 97
Arthropods, 16, 77
Arvicanthis, 112
Asia, 63, 106
Asters, 131
Attenuation, 73, 114, 153

Australia, 60, 87, 91, 94, 122, 141, 151–155, 165, 171, 196
Austria, 59

Bacteria, 3, 4, 6, 195
Bahrein, 121
Balance, 5
Balfour, A., 100
Balkins, 113
Baltimore, 98
Bandicoot, 200
Bang, O., 181
Baron, S., 179
Bats, 83, 109, 143–145
Bawden, F., 22
Bechuanaland, 106
Bedson, S.P., 195
Beetles, 125, 130
Belem, 114
Belfast, 60
Belgium, 156
Beneficent viruses, 130–132
Biggs, P.M., 185
Bile-salts, 12
Binary fission, 21, 195
Birds, wild, 109, 197
Bittner, J.J., 187
Black, F.L., 22
Black Sea, 13
Black-crowned night herons *v.* Herons
Blesbuck, 115
Blind alley, 80, 110, 144
Blind passage, 9
Blindness, 198
Bobcats, 145
Bokhara, 54
Bolivia, 97, 100
Boots, 134
Boston, 38, 54, 98, 177
Brazil, 114
Britain, 59, 120, 122, 141, 144, 156, 198, 205

Browsing on cells, 26
Buckland, F.E., 47
Budgerigars, 196
Buescher, E.L., 88, 177
Bugs, 78, 124
Bulgaria, 112, 121
Burmester, B.R., 182
Burnet, F.M., 22, 26, 95, 175
Burst of cell, 28
Bush-babies, 103
Butterflies, 123–124, 176
Buzzards, 160

Cairo, 71
Canada, 79–80, 112, 125, 171, 192
Canberra, 152
Capsids, 12
Capsomeres, 12, 14, 115
Carcinogenic agents, 181, 187
Carmichael, E.A., 163
Caribbean, 47, 80, 87, 93–94, 98
Carriers, 4
Caterpillars, 124, 202
Cats, 141, 146, 190
Cattle, 115, 121, 146, 190, 200
Celery, 131
Ceratopogonidae, 113
Cercopithecus, 103
Chemotherapy, 205
Chicago, 43, 45
Chickens, v. fowls
Children, infection in, 8, 42, 164
Children's parties, 42, 48
Chile, 149, 156
Chilling, 45
Chimpanzees, 72, 173
China, 61
Chipmunks, 112
Chloroform, 12
Chrysops, 113
Cicadellids, 128–129
Cicadulina, 129
Cincinnati, 5
Circulatory transmission, 126
Civet-cats, 95, 142
Classification, 12, 207
Cloud-babies, 50
Coccids, 130
Cold (chilling), 45
Colias, 124
Colladonus, 131
Colobus, 103
Colorado, 112
Common cold research unit, 42
Complement-fixation, 37
Congenital defects, 179
Congenital infection (transmission), 173–
    179

Convergent evolution, 13
Coons, A.H., 38
Copenhagen, 58
Cornea, 205
Costa Rica, 102
Cottontail rabbits, 149, 186
Coyotes, 144, 146
Crane-flies, 14
Crashes in population, 145
Crimea, 112
Crowding, 8, 47
Crystals, 30
Cuba, 99
Culex, 77, 84–85, 87–92, 120, 152, 155
Culicoides, 81, 113, 115
Culiseta, 80
Cyprus, 15
Cytopathic (cytopathogenic) effect, 15, 24
Cytoplasm, 21
Czechoslovakia, 76

Dagger-worm, 135
Dalbulus, 131
Damaliscus, 115
Darling river, 91
Dasypterus, 144
DDT, 113
Deer, 107, 118
De Lesseps, F., 99
Delphacodes, 131
Denmark, 59–60, 181
Deoxycholic acid, 12
Deoxyribonucleic acid, 11, 18–21, 186
Dermacentor, 112
Dermanyssus, 84
Desmodus, 143
Dhows, 105
Diprion, 117
Disease from upset balance, 6
DNA, v. Dioxyribonucleic acid
Dodder, 5, 137
Dogs, 18, 117, 141, 146, 173
Donkey, 116
Dowling, H.F., 43, 45
Droplets, 48–49
Ducks, 109, 183, 197
Dust, 49, 174

Eclipse phase, 27
Ecology, 10, 24, 202
Eelworms, v. Nematodes
Eggs, growth in, 9, 20, 57
Egypt, 87, 198
Elephant-grass, 106
Ellerman, V., 181
Elution, 26
Embryonic tissue, 25
Enders, J.F., 15, 24

Entebbe, 81
Entry into cell, 26–27
Envelope, *v.* membrane
Ether, 12, 20
Ethiopia, 7, 105
Evolution, 4, 22, 149, 154–155
Eyes, 49, 198–199

Faeces, 68, 71
Faëroes, 7, 172, 197
Fatigue and susceptibility, 69
Fenner, F., 152–155, 159
Ferret, 56–57, 117
Fiji, 7, 172
Filaments, 29
Finlay, C., 98
Flea-beetles, 130
Fleas, 156–160, 199–200
Flies, 72, 123, 130
Flooding, 91
Florida, 112
Fluorescent antibody, 27, 38
Formalin, 204
Fowls, 173, 180–185
Foxes, 107, 142–144, 160
France, 123, 156
Frogs, 192
Fukushi, 129
Fulmars, 197
Funambulus, 111
Fungi, 133, 136

Galago, 103
Gamma-globulin, 179
Ganglia, 166
Garter-snake, 83
Gel-diffusion, 38
Germany, 60, 123, 153, 170, 198
Gigantolaelaps, 118
Gipsy-moth, 176
Globulins, 34
Glossina, 113
Goats, 110
Goose-cells, 37
Gorgas, W.C., 99
Gorgon, 148
Grapes, 136
Grasshoppers, 130
Great Lakes, 51
Greece, 94, 121
Green, R.G., 22
Green-Laidlaw hypothesis, 22
Greenwood, M., 54
Gregg, N.S., 178
Grivet, 103
Gross, L., 173, 188–190
Ground-nuts, 128
Ground-squirrels, 112

Grouse, 108
Guatemala, 102
Guinea-pigs, 123, 173

Habel, K., 191
Habitual herpetikers, 163
Haemadsorption, 17, 37
Haemagglutinin (-ination), 17, 26, 37, 57, 115
Haemagglutinin-inhibition, 37, 57
Haemagogus, 100
Haemaphysalis, 111, 200
Haematopinus, 120
Haemophilus, 65
Hamre, D., 43
Hamster, 173, 184, 190–192
Hanafusa, H., 184
Handkerchiefs, 49
Hares, 149, 160–161
Harmless viruses, 6, 132
Heads of bacteriophage, 26
Heart defects, 178
Heating viruses, 168, 171
Hedgehogs, 123
HeLa cells, 25
Helical virus, 12, 17
Helper virus, 184
Heron, 87–90
Hibernation, 83
Hirst, G.K., 57
Holland, 42–43, 47, 61, 153
Honduras, 102
Hong Kong, 61
Hope-Simpson, R.E., 46, 166
Hoppers, *v.* Leaf-hoppers
Horses, 79, 116, 170
Host-components, 29
Howler monkeys, 100
Huebner, R.J., 18
Hull, R.N., 26
Humidity, 46–47
Hungary, 147
Husbandry, 128, 204
Hybridization, 28
Hygiene, 70, 198, 202
Hymenoptera, 123

Iceland, 59, 76, 170–172, 202
Icosahedron, 14, 21
Illinois, 51
Immunity, 34–36
Immunological barrier, 91
Immunological tolerance, 175
Impala, 121
Inclusion body, 19, 21, 30, 174
Incomplete virus, 27
Incubation period, 78, 93, 168
India, 64, 87, 90, 97, 108, 130, 142, 205

Induction of phage, 29
Inefficiency, of viruses, 52
Insects, effect of viruses on, 123–124, 176–177
Insects and virus origin, 22, 199
Interferon and interference, 36, 74, 162, 179, 205
Intrinsic incubation period, 78
Introduction of new viruses, 7
Iowa, 9, 66
Irak, 121
Iran, 142
Iceland, 59, 64
Isaacs, I., 179
Isotherms, 94
Israel, 88, 115, 121
Italy, 76, 113, 156, 161, 200
Itching, 146
Ixodes, 107–111

Jackal, 142
Jackson, G.G., 43, 45
Jamaica, 83
Jaws, tumours of, 193
Japan, 8, 88–90, 94, 112
Johnson, H.N., 142, 145
Johnson, R.T., 165
Jugoslavia, 144
Jungle, 95, 100, 202
Jungle-fowl, 111

Kansas, 134
Karakul sheep, 170
Kent, 156
King Edward potatoes, 133
Koprowski, H., 72
Korea, 97
Kweichow, 61

Laboratory infections, 201
Laidlaw, P.P., 22
Lancashire, 120
Langur, 110
Larvae, *v.* caterpillars
Lasimus, 144
Latent infections, 9, 120, 162–172
Leaf-hoppers, 127–129
Leopard-frog, 192
Lepidoptera, 123
Lepus, 149, 160
Lidwell, O.M., 42, 45
Link-hosts, 89
Lipids, 12
Liverpool, 60
Lizards, 109
London, 42, 58
Louse, 199–200

Lung–worms, 64–65
Lymph-nodes, 32
Lymphoid tissue, 68
Lysogeny, 29

Macacus (Macaques), 95
Madagascar, 76
Madras, 142
Magrassi, F., 59
Maintenance hosts, 78
Maitland, H.B., 24
Maize, 129
Malaya, 90, 95, 108, 111
Malnutrition, 164
Maramorosch, K., 129–132
Markers, 73
Marmosets, 19, 100
Marmota, 151
Maternal antibody, 108
Mealy-bugs, 128, 130
Mechanical transmission, 119, 133
Mediterranean, 93, 168
Melilotus, 129
Melopsittacus, 196
Membranes, 20
Metacheirus, 102
Mexico, 75, 121
Meyer, K.F., 196–197
Mice, infant, 75, 173
Mice, diseases of, 173–174
Michigan, 82
Migrating birds, 82–83, 122
Miles, J.A.R., 196
Milk, 110, 188, 200
Milkmaids, 119
Miller, C.P., 10
Mink, 146
Mites, 83, 97, 119, 130, 199–200
Moles, 173
Mongoose, 142–143
Monkeys, infection in wild, 19, 95 100–104, 110–111
Monkeys kidneys, 19, 25, 75, 173, 177, 191
Montevideo, 150
Mortality, 66, 153
Mosquitoes, 77, 150
Moths, 123, 130, 176
Mucin (mucoprotein), 17
Mucous membranes, 35
Mulder, J., 63
Mules, 116
Muramic acid, 21, 195
Murray river, 91, 157
Muscidae, 113
Musk-rats, 110
Mustelidae, 142, 145
Mykytowycz, R., 153
Myzus persicae, 127

National Institutes of Health, 18
Nematodes, 135
Neodiprion, 124
Nephotettix, 131
Nerves, virus-transport in, 32, 164
Netherlands, *v.* Holland
Neutralization, 34, 43
Neva, 177
New Guinea, 91
New Jersey, 66, 80, 144
New South Wales, 91, 152
New York, 57, 64, 98
New Zealand, 122, 156, 171
Newcastle, 42
Nicaragua, 102
Nose, 41
No-see-ums, 115
Noguchi, H., 9
Nuba mountains, 105
Nucleic acids, 11, 14, 23, 28
Nucleic acids, infectious, 28, 135
Nucleoprotein, 12, 28
Nucleus, 21, 27–29
Nye, R.N., 24

Old people, 34, 63
Olpidium, 136
Ontario, 112
Opossums, 102, 145, 173
Organ culture, 25
Origin of viruses, 22
Orphan viruses, 15
Oryctolagus, 149–150, 186
Oryzomys, 118
Overwintering, 82

Pakistan, 115
Panama, 99, 117
Parakeets, 195–196
Parker, F., 24
Parkman, P.D., 177
Parrots, 195–196
Passenger viruses, 181, 190
Payne, L.N., 185
Peromyscus, 84
Persistent transmission, 126
Pfeiffer, R., 56, 65
Phagocytes, 26
Pheasants, 7, 183
Philippines, 94, 96
Phlebotomus, 113–115
Phloem, 33
Phobey cats, 143
Pigs, 68, 89, 146, 173, 179
Pig-louse, 120
Pigeons, 87, 197
Placenta, 108, 175, 200
Plaques, 25

Plastic plates, 37
Plymouth rocks, 183
Polar explorers, 4, 46
Polyhedra, 123, 176
Portugal, 98, 115, 147
Potatoes, 126, 133
Precipitins, 37
Presbytis, 110
Propagative transmission, 126
Prophage, 29
Protein, 11, 27
Provoking factors, 66
Psittacines, 196
Psychodidae, 113
Pure cultures, 7
Purines, 11
Pyrimidines, 11

Queensland, 91, 200

Rabbits, 9, 149–160
Raccoons, 145
Rana, 192
Rashes, 76, 120, 165, 177
Rats, 147, 173, 184, 199
Receptors, 26
Recombinants, 28
Recruits, 50
Red blood cells, 26
Red-tail monkeys, 103
Reed, W., 99
Reeves, W.C., 83
Release from cell, 28–29
Resistance, acquired, 34
Resistance, genetic, 155
Rhineland, 161
Rhinocoris, 124
Rhipicephalus, 112
Ribonuclease, 28
Ribonucleic acid, 11, 14–17, 129
Rivers, T.M., 9
RNA, *v.* Ribonucleic acid
Robbins, F.C., 24
Robins, 124
Rockefeller Foundation, 99, 114
    Institute, 9
Rodents, 83, 107, 109–111
Roller, drum, 25
Roots, 134
Rothschild, M., 159
Rous, P., 181, 186
Rowe, W.P., 18, 174
Rubin, H., 184
Rye, 132

Sabethes, 102
Sabin, A.B., 72
Sage-brush rabbit, 149, 157

Salamander, 193
Salisbury, 42, 45, 47
Saliva, 141, 144, 174
Salivary glands, 173, 189
Salk, J., 72
Sanarelli, G., 150
Sandflies, *v.* Phlebotomus
Sardinia, 59
Satellite viruses, 136
Sawflies, 123–124, 177, 202
Scandinavia, 60, 110, 123
Scherer, W.F., 88
Scotland, 8, 108, 126, 171
Seagulls, 123
Season, effect of, 47
Sentinels, 77, 103
Serotherapy, 205
Sheep, 108, 115, 148, 170–173, 200
Sheep-ked, 199
Shope, R.E., 9, 64, 149, 186
Shrew, 111
Siberia, 80, 90, 110
Simulium, 113
Singapore, 61
Skunks, 142–143, 145
Slaughter policy, 122, 170, 203
Smith, K.M., 176
Snakes, 83
Sneeze, 4, 48
Soil, transmission in, 134
South Africa, *v.* Africa
Spain, 59, 98, 115, 147, 156
Sparrow, 7
Sperm, 183
Sphinx, 176
Spider monkey, 100
Spikes, 17
Spilopsyllus, 158, 160
Spitzbergen, 44
Squirrel, 95, 111, 151
Stable-fly, 116, 170
Starling, 122
Stomoxys, 116–117, 170
Strawberries, 136
Stress, 8, 45, 51–52
Stylets, 27, 126
Sudan, 105
Sugar-beet, 128, 131
Sulphonamides, 3
Swan, C., 178
Sweden, 59, 69, 71, 73, 179
Swift, H.F., 10
Swill, 122
Swimming-bath, 199
Swine, *v.* pig
Switzerland, 59, 156
Symbionts of insects, 22
Sylvilagus, 149, 186

Symmetry, 12
Syncytia, 50
Syringes, 169, 204

Tabanidae, 113, 116–117, 144, 170
Tadarida, 144
Tanganyika, 85
Tapetis, 149
Target organ, 30, 41
Tasmania, 196
Taxonomists, 13
Temperature, effects of, 45, 47, 93
Thailand, 85, 94, 96
Thamnophis, 83
Thrips, 130
Ticks, 107–112, 159, 199–200
Tierra del Fuego, 153, 156
Tillett, W.S., 9
Tipula, 14
Tissue culture, 24–26
Tobacco, 134
Togoland, 77
Tokyo, 88
Toleration (and *v.* immunological toler-
    ance), 5
Tonsils, 18
Towers in jungle, 81
Tractors, 134
Transfusion, 168
Transmission, 41
Transovarial, 109, 173
Traub, E., 175
Trichodorus, 135
Trinidad, 77, 82–83, 92, 100, 102, 118, 143
Turkeys (bird), 183, 197
Turkey (country), 115, 121, 144
Turtle, 83
Tyrrell, D.A.J., 47

Uganda, 81, 85, 112
United States, 62, 75, 79, 94, 120, 129, 143,
    174, 198
Uridine, 205
Uruguay, 150
USSR, 73, 97, 113
Uzbekistan, 112

Vaccines as stress, 51
    influenza, 58, 63, 205
    poliomyelitis, 72
    smallpox, 74
    yellow fever, 74, 204
    other viruses, 204–205
Vampires, 143–144
van Loghem, J.J., 43
Vectors, 4
Vegetation changes, 160–161

Venereal infection, 198
Vermont, 192
Vertical transmission, 173
Vesicles, 163, 165
Victoria, 91, 152
Viraemia, 79, 83, 168
Virulence, changes in, 153, 158–159, 175
Viverridae, 142, 145
Voles, 142, 160

Wales, 64, 98
Wart-hog, 7
Washington, 18
Weasels, 142
Weather, 47
Weller, T.H., 24, 177
West Indies *v.* Caribbean
White fly, 130, 133
WHO, 58–59, 106, 202–203

Wildebeest, 7, 148
Williams, R.E.O., 43, 45
Winds, 128
Winter, 47
Wolves, 142
Woodchucks, 157
World Influenza Centre, 58
Worms, 64–66, 113, 133

Xenopsylla, 200
Xiphinema, 135

Yolk, antibody in, 91
Yolk-sac, 20

Zinsser, H., 199
Zoonases, 7
Zoospores and zoosporangia, 136
Zygodontomys, 118

Index of Virus Diseases, Viruses and Other Disease-agents

(Main references in heavy type)

Abortion in mares, 19
Adenovirus, 13, 14, **18**, 24, 28, 30, 35, 41, **50–51**, 74
African horse-sickness, 81, 115, **116–117**, 202
African swine fever, 7, **147**, 202
Anaemia, equine, 35, **170**
Arabis mosaic, 136
Arbovirus, **16**, 32, 35, 77, 207
Aster yellows, 129, **131–132**
Australian X-disease, 90–91

Bacteriophage, 4, 11, 14, 25–26, 28–29
Bedsonia, 195
Bittner's virus, 187–188
Blue-tongue, 81, **115–116**, 179, 202
Bornholm disease, 15
Bronchitis, in children, 50
Burkitt tumour, 193–194
B-virus of monkeys, **19**, 203

Cacao, swollen shoot, 128, 130
Cancer (carcinoma), 9, 180, 190
Cattle-plague, 13, 17, 167
Central European tick-borne fever, 109
Chickenpox, 9, **165–167**
Chikungunya, **85–86**, 96
Chlamydozoon (-ozoäceae), 195–199
Choriomeningitis, 174–176
Clover club-leaf, 129
Cocal, 118
Cold (common cold), 16, 30, 34, **41–53**
Colorado tick-fever, 112, 200
Conjunctivitis, 18, 41, 195, 198–199
Contagious pustular dermatitis, 121
Corky areas in potato, 134
Cowpox, 20, 119
Coxsackie, 13, 15–16, 49–50, 68, 74, **75–76**
Cucumber mosaic, 134

Curly-top of beet, 129
Cytomegalovirus, 19, **173–174**

Dengue, 85, **93–97**, 114
Deoxyvirus, 180, 207
Dog distemper, 13, 17, 167

Echo, 15–16, 24, 35, 68, 74, **76**
Ectromelia, 13, 120
Encephalitis, cattle, 195
  equine, 7, **79–84**, 119, 202
  tick-borne, **108–110**, 119
  other, 8, 79, 87, *v.* Murray Valley, St Louis, etc.
Enterovirus, 6, 15, 41, **68–76**
Erythroblastosis, 182

Fan-leaf of grapes, 135–136
Far East tick-borne encephalitis, 108, 110
Fibroma of hares, 161
  rabbits, 149–151
  squirrels, 151
Foot and mouth disease, 13, 14, 16, 35, 117
Frog tumours, 192–193, **121–123**, 203–204
Fowl paralysis, 185
  plague, 27
  pox, 120
  tumours, 18, 35, **180–185**

German measles, *v.* Rubella

Haemorrhagic fever, 93, **96–97**
Hepatitis canine, 13, 18
  in horses, 170
  of man, 21, 35, **167–170**, 204
Herpes, 8, 13, 19, 29, 32, 35, 52, **162–165**, 205
Herpes virus group, 18, **19**, 27, 30
Herpes zoster, 165–167
Hog cholera, *v.* swine fever
Hydrophobia, *v.* rabies

Iccaraci, 114
Infectious chlorosis, 130
Influenza (and influenza A), 17, 26, 29, 34–35, 46–47, **54–67**, 205
   Asian, 59, **61–64**
   B, 17, 50, 57, 66
   C, 17, 57, 66
   swine, 9, **64–66**
Insect pathogens, 2, **123–124**, 176–177, 202

Jaagsiekte, 171
Japanese encephalitis, 8, **87–92**, 106

K-virus, 19
Kyasanur forest, 97, 108, **110–111**, 203

Langat, 108, 111
Laryngotracheitis, 19
Lettuce big-vein, 136
Leukaemia, mouse, 10, 18, 173
   fowl, 18, **181–184**
Louping-ill, 8, **108–109**, 171
Lymphocytic choriomeningitis, 174–176
Lymphogranuloma, 195, **198**
Lymphoma (-omatosis), 182–183

Mad itch, 13, 19
Maedi, 170
Maize streak, 129
Malignant catarrh, 7, **147–148**, 202
Marek's disease, **185**
Measles, 7, 13, 17, 34, 172
Mite-borne typhus, 200
Miyagawanella, 195
Molluscum contagiosun, 119
Monkey tumour, 192
   -viruses, 26, 75
Mumps, 13, 30, 34
Murray Valley fever, 80, 87, **90–91**, 106
Myeloblastosis, 182
Myxoma (-omatosis), 5, 20, 25, 120, **149–161**, 202
Myxoviruses, **17**, 26, 29, 149

Nairobi sheep disease, 112
Necrotic spots in hyacinths, 134
Neurolymphomatosis, 182, **185**
Newcastle, disease, 13

Oat mosaic, 131
Omsk haemorrhagic fever, 97, 108, **110**
O'nyong-nyong, 85–86
Orf, 120
Ornithosis, 196–198

Papilloma, rabbit, 19, 123, **186–187**
   other species, 14, 19, 30, 123
Papovavirus group, 18
Paracrinkle, 133

Para-influenza, 17, 41, 46, **49**
Parvovirus, 14
Phlebotomus fever, 113–115
Plant-viruses, 125–137
Pleurodynia, 15
Pneumonia, 54, 195
Poliomyelitis, 8, 14–16, 24–25, 32, 47, **68–75**, 119, 143, 202–3
Polyhedrosis, 21, 123–124, 176
Polyoma, 19, 191
Potato leaf-roll, 127
Poxes of various species, 20, 119–121
Poxvirus group, 18, **20**, 29–30, 34, 37
Powassan, 108, 112
Pseudorabies, 13, 19, **146–147**
Psittacosis, 21, 195, **196–198**
Pulmonary adenomatosis, 171

Q-fever, 200

Rabies, 5, 18, 32, **141–146**, 203, 205, 207
Rabbit, latent virus, 9
   myxoma, 149
Reovirus, 14, **16**, 207
Respiratory syncytial, 17, **50**
Rheumatic fever, 10
Rhinopneumonitis, 19
Rhinotracheitis, 19, 198
Rhinovirus, 16, 35, **41–53**, 205
Ribovirus, 14, 207
Rice dwarf, 131
   stunt, 129
Rickettsia, 21, 108, **199–201**
Rida, 171
Rinderpest, *v.* cattle-plague
Ring spot of raspberry, 135
Rocky mountain spotted fever, 200
Rosette of ground-nuts, 128
Rous sarcoma, *v.* fowl tumour
Rubella, 57, **177–179**
Russian spring-summer encephalitis, 108

St Louis encephalitis, 79, 87, **91–92**, 106
Salivary viruses, 173
Sandfly fever, 113–115
Sarcoma, 161, 181
Scrapie, 171
Scrub typhus, 200
Semliki forest, 86
Shingles, 165–167
Simian *v.* monkey
Simian virus 40 (SV40), 19, 75, 191
Sleeping-sickness, 79, 113
Smallpox, 20, 32, 74, **119–120**, 204–205
Swine fever, 147, 179
   flu, 9
   pox, 120
Swollen shoot of cacao, 128, 130

Teschen disease, 16, 76
Tensaw, 83, 85
Tick-borne encephalitis, **108–112**, 119
  fever, 108
Tipula, 14, 22
Tobacco mosaic, 125, 130, **133–134**
  necrosis, 134, 136
    rattle, 135
    spotted wilt, 130
    stunt, 136
Trachoma, 21, 195, **198–199**
Trypanosome, 79, 113
Tuberculosis, 172
Tumour-viruses, **180–194**
Turnip-viruses, 130
Typhus, 2, **199–200**, 202

Uzbekistan haemorrhagic fever, 112

Vaccinia, 20, 24, 119, 205

Vacuolating virus, *v.* simian virus, 40
Varicella, 6, 9, 165
Variola, *v.* smallpox
Venezuelan equine encephalomyelitis, 80
Vesicular stomatitis, **117–118**
Visna, **171–172**

Warts, *v.* papilloma
West Nile, **87–88**, 106
Western equine, *v.* encephalitis
Western X, 131
Wheat mosaic, 131, 134
Wound tumour, 129, 134

Yaba, 192
Yellow-fever, 7, 12, 34, 74, **98–106**, 203–204
Yellows of beet, 5, 128

Zoster, **165–167**

NO 1 0 '87			

DEMCO